Martin Gebel

Multivariate calibration of classifier scores into probability space

Martin Gebel

Multivariate calibration of classifier scores into probability space

Comparison of uni- and multivariate calibration techniques for classification and introduction of the Dirichlet Calibration

Südwestdeutscher Verlag für Hochschulschriften

Impressum/Imprint (nur für Deutschland/ only for Germany)
Bibliografische Information der Deutschen Nationalbibliothek: Die Deutsche Nationalbibliothek verzeichnet diese Publikation in der Deutschen Nationalbibliografie; detaillierte bibliografische Daten sind im Internet über http://dnb.d-nb.de abrufbar.

Alle in diesem Buch genannten Marken und Produktnamen unterliegen warenzeichen-, marken- oder patentrechtlichem Schutz bzw. sind Warenzeichen oder eingetragene Warenzeichen der jeweiligen Inhaber. Die Wiedergabe von Marken, Produktnamen, Gebrauchsnamen, Handelsnamen, Warenbezeichnungen u.s.w. in diesem Werk berechtigt auch ohne besondere Kennzeichnung nicht zu der Annahme, dass solche Namen im Sinne der Warenzeichen- und Markenschutzgesetzgebung als frei zu betrachten wären und daher von jedermann benutzt werden dürften.

Verlag: Südwestdeutscher Verlag für Hochschulschriften Aktiengesellschaft & Co. KG
Dudweiler Landstr. 99, 66123 Saarbrücken, Deutschland
Telefon +49 681 37 20 271-1, Telefax +49 681 37 20 271-0
Email: info@svh-verlag.de
Zugl.: Dortmund, TU, Diss., 2009

Herstellung in Deutschland:
Schaltungsdienst Lange o.H.G., Berlin
Books on Demand GmbH, Norderstedt
Reha GmbH, Saarbrücken
Amazon Distribution GmbH, Leipzig
ISBN: 978-3-8381-1224-4

Imprint (only for USA, GB)
Bibliographic information published by the Deutsche Nationalbibliothek: The Deutsche Nationalbibliothek lists this publication in the Deutsche Nationalbibliografie; detailed bibliographic data are available in the Internet at http://dnb.d-nb.de.

Any brand names and product names mentioned in this book are subject to trademark, brand or patent protection and are trademarks or registered trademarks of their respective holders. The use of brand names, product names, common names, trade names, product descriptions etc. even without a particular marking in this works is in no way to be construed to mean that such names may be regarded as unrestricted in respect of trademark and brand protection legislation and could thus be used by anyone.

Publisher: Südwestdeutscher Verlag für Hochschulschriften Aktiengesellschaft & Co. KG
Dudweiler Landstr. 99, 66123 Saarbrücken, Germany
Phone +49 681 37 20 271-1, Fax +49 681 37 20 271-0
Email: info@svh-verlag.de

Printed in the U.S.A.
Printed in the U.K. by (see last page)
ISBN: 978-3-8381-1224-4

Copyright © 2010 by the author and Südwestdeutscher Verlag für Hochschulschriften Aktiengesellschaft & Co. KG and licensors
All rights reserved. Saarbrücken 2010

Contents

1. **Introduction**   3
   1.1. Basic Notation 6
2. **Classification with supervised learning**   9
   2.1. Basic Definitions 9
      2.1.1. Decision–theoretical groundwork 10
      2.1.2. Classification problems and rules 12
   2.2. Post–processing with Membership values 14
      2.2.1. Comparison and combination of multiple classifiers 15
      2.2.2. Cost–sensitive decisions 15
      2.2.3. Multi–class problems 16
   2.3. The aim of calibration 16
   2.4. Performance of classification rules 18
      2.4.1. Goodness estimates 20
      2.4.2. Estimating performance reliably 21
      2.4.3. Learning of rules and the problem of overfitting 23
      2.4.4. Further Performance Measures 24
3. **Regularization methods for classification**   27
   3.1. Support Vector Machine 28
      3.1.1. Separating Hyperplanes for two classes 28
      3.1.2. Generalization into high–dimensional space 33
      3.1.3. Optimization of the SVM Model 34
   3.2. Artificial Neural Networks 35
      3.2.1. McCulloch–Pitts neuron and Rosenblatt perceptron 36
      3.2.2. Multi Layer Perceptron 38
      3.2.3. Generalized Delta Rule 40

	3.2.4. Regularization and grid search	41
3.3.	Consistency of convex surrogate loss functions	42

4. Univariate Calibration methods   45
    4.1. Simple Normalization 46
    4.2. Calibration via mapping 47
        4.2.1. Logistic Regression 48
        4.2.2. Piecewise Logistic Regression 49
        4.2.3. Isotonic Regression 51
    4.3. Calibration via Bayes Rule 53
        4.3.1. Idea 54
        4.3.2. Gaussian distribution 55
        4.3.3. Asymmetric Laplace distribution 56
    4.4. Calibration by using assignment values 60
        4.4.1. Calibration by inverting the Beta distribution 61

5. Multivariate Extensions   65
    5.1. Procedures in Multi–class classification 65
    5.2. Reduction to binary problems 69
        5.2.1. Common reduction methods 69
        5.2.2. Sparse ECOC–matrix 71
        5.2.3. Complete ECOC–matrix 74
        5.2.4. Comparison of ECOC–matrix based reduction approaches 76
    5.3. Coupling probability estimates 78
    5.4. Calibration based on the Dirichlet distribution 79
        5.4.1. Dirichlet distribution 79
        5.4.2. Dirichlet calibration 81

6. Analyzing uni– and multivariate calibrators   84
    6.1. Experiments for two–class data sets 84
        6.1.1. Two–class data sets 85
        6.1.2. Results for calibrating classifier scores 87
        6.1.3. Results for re–calibrating membership probabilities 92
    6.2. Experiments for multi–class data sets 97
        6.2.1. K–class data sets 98

      6.2.2. Calibrating $L2$–SVM scores in multi–class situations . . . 100
      6.2.3. Calibrating ANN–classification outcomes for $K > 2$ 105

7. Conclusion and Outlook   111

A. Appendix   122
    A.1. Distributions 122
      A.1.1. Gamma distribution 122
      A.1.2. Beta distribution 122
    A.2. Further analyses of calibration results 124

Acknowledgements

Firstly, I would like to thank my advisor Prof. Claus Weihs for motivating and supporting me in finishing this thesis, especially in the time after I left university. Secondly, I would like to thank the **Deutsche Forschungsgemeinschaft (DFG)** and the Faculty of Statistics at the University of Dortmund for making the Graduiertenkolleg and hence my project possible.

Furthermore, there are several people who supported me during the work on this thesis:

- Gero Szepannek and Karsten Lübke for fruitful discussions and help about classification and Machine Learning,

- Uwe Ligges for support in R and UNIX,

- Sabine Bell and Eva Brune for all administrative support,

- Irina Czogiel, Verena Kienel, Tina Müller and Rita Mirbach for proof–reading and

- all my colleagues and friends at the University of Dortmund, especially Irina Czogiel, Carsten Fürst, Tina Müller and Ana Moya who made the time at the university an enjoyable time.

Apart from that, I would like to thank my parents Doris and Manfred Gebel for all their support over all these years.

Finally, my gratitude belongs to my beloved wife Rita Mirbach for her support, patience, motivation and understanding.

1. Introduction

Data analysis is the science of extracting useful information from large data sets or databases. In the computer era, databases and hence analysis of data have attained more and more interest, in particular classification with supervised learning as an important tool with various application fields. Classification can be used for speech and image recognition, in medical diagnosis as well as for the determination of credit risks, just to name a few.

Thus, the number of competing classification techniques is rising steadily, especially since researchers of two different areas, Statistics and Computer Science, are working on this topic. Nevertheless, standard methods such as the Discriminant Analysis by Fisher (1936) and the Support Vector Machine by Vapnik (2000) are still the main classifiers being used. The common characteristic of all classifiers is that they construct a rule for the assignment of observations to classes. For each observation such classification rules calculate per corresponding class membership values. The basis for rule construction is a given sample of observations and corresponding class labels determined by a supervisor. Future observations with unknown class labels are assigned to the class that attains the highest membership value.

The methods for classification can be separated into two competing ideologies, Statistical Classification and Machine Learning, which have their origins in Statistics and in Computer Science, respectively. Membership values supplied by statistical classifiers such as the Discriminant Analysis or certain Machine Learners are *membership probabilities* which reflect the probabilistic confidence that an observation belongs to a particular class. Other Machine Learners, e. g. the Support Vector Machine or Artificial Neural Networks, only generate unnormalized scores. In contrast to the membership probabilities, the unnormalized scores do not reflect the assessment uncertainty and neither sum up to one nor lie in the interval $[0, 1]$.

Comparing these two kinds of membership values reveals several advantages of probabilistic membership values. Membership probabilities make classifiers and their results comparable. This is desirable because of the high number of classification techniques. Furthermore, probabilities are useful in post–processing of classification results, i. e. the combination of membership values with a given cost matrix. Consequently, a unifying framework for supplying probabilistic membership values and hence for the comparison of classification methods is an important target. Such a framework the calibration process offers in transforming membership values into probabilities which cover the assessment uncertainty.

Moreover, even membership values which claim to cover the assessment uncertainty can be inappropriate estimates. The Naive Bayes classifier and Tree procedures yield membership values which tend to be too extreme, as shown in analyses by Domingos & Pazzani (1996) for Naive Bayes and by Zadrozny & Elkan (2001b) for Tree. Therefore, membership values generated by these classifiers may need to be re–calibrated in some situations as well.

In recent years, calibration of membership values has attained increased attention by researchers. Different ways of calibrating classifiers have been introduced, among them Logistic Regression by Platt (1999) as the most highly approved approach. This calibrator was later extended by Zhang & Yang (2004) into a locally operating technique using Piecewise Logistic Regression. Additionally, another calibration method was introduced by Zadrozny & Elkan (2002) by incorporating Isotonic Regression. Common characteristic of all these methods is that they are only applicable for the two–class case, since they estimate a probability for one class by mapping the membership value to a calibrated probability and estimate the membership probability for the remaining class with the complement.

However, two further calibrators exist which do not map membership values directly to calibrated probabilities but split membership values into partitions before calibration. Firstly, there is the Bayesian method by Bennett (2003) which splits membership values according to their true class. Secondly, a method using the inverted Beta distribution function and a partition of membership values according to their assigned class was introduced by Garczarek (2002). Here, the calibration function is learnt independently in each partition.

In polychotomous classification tasks, these univariate calibration procedures are especially used for Support Vector Machines and Artificial Neural Networks, since these methods usually reduce multi–class tasks to binary decisions and give no probabilistic output. Thus, the membership values generated for the binary decisions are calibrated to membership probabilities. Afterwards, these two–class probability vectors are usually combined with the pairwise coupling algorithm by Hastie & Tibshirani (1998) to one probability matrix considering the K–class task. This thesis introduces an alternative direct multivariate calibration method based on probability theory for K–class regularization methods. In just one step, this alternative method generates the probability matrix of membership probabilities as a realized set of a Dirichlet distributed random vector. This flexible method has got the advantage that it is on the hand applicable to binary outcomes generated by any reduction approach and on the other hand is also directly applicable to multi–class membership values.

Chapter 2 provides the groundwork for the classification problem covering the decision theory and its application to the learning of classification rules. Reasons for a preference of membership probabilities to unnormalized scores as well as the desired output of a calibrator are also obtained in this chapter. Additionally, performance measures and estimation procedures for these measures are presented. This includes the introduction of the *Well–Calibration Ratio* based on the concept of Well–Calibration by DeGroot & Fienberg (1983).

Chapter 3 introduces the two standard regularization methods for classification, Artificial Neural Networks and Support Vector Machine, and how these methods calculate membership values. Moreover, connections between the choice of the loss function in regularization and the ability to estimate membership probabilities are presented here.

Chapter 4 gives an overview of the currently known univariate calibration methods which can be divided into four different groups, simple normalization as well as calibration via mapping, by using Bayes' rule and usage of assignment values. Multivariate extensions are introduced in Chapter 5 including the standard procedure of reduction to binary classification tasks with subsequent pairwise coupling and a new calibrator based on the Dirichlet distribution. With this Dirichlet calibration method the output of the binary reduction algorithms are transformed into Beta distributed random variables and afterwards combined to realizations of a Dirichlet distributed random vector. This calibrator is furthermore applicable for a multi–class probability matrix with regarding the columns as Beta distributed random variables and a similar combination to a matrix of Dirichlet distributed probabilities.

In Chapter 6, the results of both an analysis for univariate and multivariate calibration methods are shown. Basis for these analyses is a 10–fold cross–validation, so that reliable performance measures are supplied. Finally, Chapter 7 gives concluding remarks and presents an outlook.

1.1. Basic Notation

In the following the notation which will be used throughout this thesis is shown, see Tables 1.1 to 1.3. This list is presented as easy reference.
Table 1.4 presents the abbreviations used in this thesis.

Table 1.1.: Notation – Vectors and matrices

Quantity	Notation	Comment
Vector	$\vec{x}$	arrow
Matrix	$\mathbf{X}$	bold typeface
Scalar product	$\langle \vec{x} \rangle$	

Table 1.2.: Notation – Sets and spaces

Quantity	Notation	Comment
Real–valued set	$\{a,b\}$	curly brace
Integer set	$[a,b]$	closed squared brackets
Excluding set	$]a,b[$	open brackets
Positive integer space	$\mathbb{N}$	
Positive real–valued space	$\mathbb{R}_0^+$	
Closure	$\bar{\mathbb{R}} := \mathbb{R} \cup \{-\infty, \infty\}$	R–bar
Hilbert Space	$\mathcal{H}$	
Training Set	$\mathcal{T}$	see Definition 2.1.7
Set size	$\|\;\|$	

Table 1.3.: Notation – Functions and operators

Quantity	Notation	Comment
Indicator function	$\mathbf{I}\,()$	
Loss function	$L\,()$	see Definition 2.1.6
Risk function	$R_L\,()$	see Definition 2.1.6
Expected value	$E\,()$	
Mean	$\bar{x}$	bar

Table 1.4.: Abbreviations

Abbreviation	Term
Ac	Accuracy
ANN	Artificial Neural Network
AS	Ability to separate
AV	Assignment Value
Cal	Calibration Measure
CART	Classification and Regression Trees
Cf	Confidence
CR	Correctness Rate
ECOC	Error–correcting output coding
IR	Isotonic Regression
LDA	Linear Discriminant Analysis
LR	Logistic Regression
MLP	Multi Layer Perceptron
NP	Non–deterministic polynomal–time
QDA	Quadratic Discriminant Analysis
PAV	Pair adjacent violators
PLR	Piecewise Logistic Regression
RKHS	Reproducing Kernel Hilbert Space
RMSE	Root Mean Squared Error
SE	Squared Error
SVM	Support Vector Machine
WCR	Well–Calibration Ratio

2. Classification with supervised learning

This chapter gives an introduction to *Classification with supervised learning*. Since classification problems are a special kind of decision problems, basic elements of decision theory are introduced in Section 2.1. Additionally, the application of Decision Theory to the classification problem is shown to supply a groundwork for the derivation of membership values by regularization–based classification rules which will be presented in Chapter 3.

Since membership values are often object to post–processing, see Section 2.2, it is preferable to generate probabilistic membership values. If membership values are not probabilistic a transformation to membership probabilities is necessary, see Section 2.3. This process is called *calibration*.

Section 2.4 gives an overview about the goodness of classification rules and introduces further performance measures. A reliable estimation of performance measures is induced by applying one of the methods presented in Section 2.4.2.

2.1. Basic Definitions

This section supplies the basic definitions of decision theory which are necessary for the solution of classification problems with supervised learning. The theory is based on Lehmann (1983) and on Berger (1985).

Moreover, the application of decision theory to classification with supervised learning is presented in this section.

2.1.1. Decision–theoretical groundwork

Statistics is concerned with the collection, i. e. observation of data, their analysis and the interpretation of the results. The problem of data collection is not considered in this thesis, since this would lead too far.

In data analysis, i. e. statistical inference, one observes p attributes and considers them as a p–dimensional vector of random variables which are determined by a true state of the world θ.

Definition 2.1.1 (Random vector)
A *random vector* of dimension p

$$\vec{X} = (X_1, \ldots, X_p)' : \Theta \to \mathbb{R}^p$$

is a map from a set of true states of the world Θ to the p–dimensional space of real values $\mathbb{R}^p$.

It also can be assumed that the random vector is not only determined by just one true state of the world θ but by a vector $\vec{\theta}$ of true states. Nevertheless, this case is omitted here, since in application of decision theory in this thesis there is only one true state of the world, i. e. the parameter *class*, to consider.

A particular realization $\vec{x} = (x_1, \ldots, x_p)'$ of the random vector is determined by an underlying joint distribution F which covers the uncertainty about the true state of the world $\theta \in \Theta$. The observed data are regarded as realizations of the random vector in a *sample*.

Definition 2.1.2
Drawing a *sample of size* N means to independently observe a random vector $\vec{X}$ N times. This leads to an observation matrix $\mathbf{X} \sim (N, p)$ where the rows $\vec{x}_i = (x_{i,1}, \ldots, x_{i,p})', i = 1, \ldots, N$, are the independently identically distributed realizations of the random vector. The set of possible outcomes $\mathcal{X} \subset \mathbb{R}^p$ is called *sample space*.

A statistical analysis has got the aim to specify a plausible estimate for the unknown value of θ. To simplify this estimation of parameter θ usually a–priori knowledge is used to narrow the underlying distribution F down to some special family of distributions $\mathcal{F}$ in terms of a parametric model:

Definition 2.1.3 (Statistical model)

A *statistical model* Λ covers the assumptions about the distribution of a random vector $\vec{X} \in \mathcal{X}$ in the form of the triple

$$\Lambda := \left(\mathcal{X}, \mathcal{A}, \mathcal{F}_{\vec{X}|\Theta}\right)$$

where $(\mathcal{X}, \mathcal{A})$ is some measurable space and $\mathcal{F}_{\vec{X}|\Theta} := \left\{F_{\vec{X}|\theta}, \theta \in \Theta\right\}$ is a corresponding family of distributions determined by parameter $\theta \in \Theta$.

Statistical decision theory supplies a general framework for the estimation of the parameter θ in which wrong estimation is penalized with a *loss function*. With the statistical model specified this framework for the decision process is build in terms of a *decision problem* in which a possible estimate is regarded as a decision d.

Definition 2.1.4 (Decision problem)

A *decision problem* consists of three items:

- True parameter $\theta \in \Theta$ with Θ as set of all possible parameters,
- Decision $d \in \mathcal{D}$ with $\mathcal{D}$ as set of all possible decisions,
- *Loss function* $L(\theta, d) : \Theta \times \mathcal{D} \to \mathbb{R}_0^+$ which determines the loss for chosen decision d when θ is the true parameter.

With the decision problem as framework and the observed sample $\mathbf{X}$ as input the estimates are derived by a *decision rule*.

Definition 2.1.5 (Decision rule)

A *decision rule* δ is a mapping function

$$\begin{aligned} \delta : \quad \mathbb{R}^p &\to \mathcal{D} \\ \vec{x} &\to d \end{aligned}$$

which assigns to every observation $\vec{x}$ a decision $d = \delta(\vec{x})$ about the true state of the world.

To compare decision rules and to find optimal rules it is required to determine the goodness of the decisions. The accuracy of a decision rule δ can be measured with the *risk function*.

Definition 2.1.6 (Risk function)
The *L–risk*

$$R_L(\theta, \delta) = E\{L[\theta, \delta(\vec{x})]\}$$

with corresponding loss function L is the expected loss resulting from the use of decision rule δ.

2.1.2. Classification problems and rules

In classification with supervised learning a sample includes not only a random vector $\vec{X}$, see Definition 2.1.1, but also an additional discrete random variable C which specifies the referring class. The class of each observation which can be for example an indicator for a disease in a clinical trial, a single letter in speech recognition or a good/bad credit risk is determined by a supervisor.

The definitions from the previous section can be applied to the classification problem with supervised learning. The aim of supervised learning is to find a rule which assigns future observations to the appropriate class. For learning the rule a sample set is required as input.

Definition 2.1.7 (Training set)
A *training set* $\mathcal{T} = \{(\vec{x}_i, c_i) : i = 1, \ldots, N\}$ is a sample set consisting of:

- An observation matrix $\mathbf{X} \sim (N, p)$ where the rows $\vec{x}_i = (x_{i,1}, \ldots, x_{i,p})'$ are the realizations of the random vector $\vec{X}$ of p *attribute variables*,

- Class vector $\vec{c} = (c_1, \ldots, c_N)'$ where element $c_i \in \mathcal{C} = \{1, \ldots, K\}$ is the corresponding realization of random class variable C.

The number of observations in the training set belonging to the particular class k, i. e. where $c_i = k$, is denoted with N_k so that $\sum_{k=1}^{K} N_k = N$.

Analogous to the decision problem, see Definition 2.1.4, the framework for the learning of decision rules for classification is the classification problem.

Definition 2.1.8 (Classification problem)
A *classification problem* consists of three items:

- True class $c \in \mathcal{C} = \{1, \ldots, K\}$,

- Assigned class $k \in \mathcal{C}$,

- Loss function $L(c, k) : \mathcal{C} \times \mathcal{C} \to \mathbb{R}$ which determines the loss for chosen class k when c is the true class.

A systematic way of predicting the class membership of an observation is a rule for classification which is the analog to the decision rule, see Definition 2.1.5.

Definition 2.1.9 (Classification rule)
A *classification rule* $\hat{c}$ is a mapping function

$$\hat{c}: \quad \mathbb{R}^p \to \mathcal{C}$$
$$\vec{x} \to k$$

which assigns for every observation $\vec{x}$ a class $k = \hat{c}(\vec{x})$.

Basis for the classification rule is usually the calculation of membership values which indicate the confidence that the realization $\vec{x}_i$ belongs to a particular class. A membership value $m_{\text{method}}(C = k | \vec{X} = \vec{x}_i)$ is produced by the classification method for every observation $\vec{x}_i$ and each class $k \in \mathcal{C}$. Hence, application of a classification method to a training set of size N leads to a membership matrix

$$\mathbf{M} := \begin{pmatrix} m_{1,1} & \cdots & m_{1,K} \\ \vdots & \ddots & \vdots \\ m_{N,1} & \cdots & m_{N,K} \end{pmatrix}$$

where rows $\vec{m}_i = (m_{i,1}, \ldots, m_{i,K})' := (m_{\text{method}}(1|\vec{x}_i), \ldots, m_{\text{method}}(K|\vec{x}_i))'$ represent the membership values for observation $\vec{x}_i$.
The classification rule

$$\hat{c}(\vec{x}_i) = \arg\max_{k \in \mathcal{C}} m_{\text{method}}(k|\vec{x}_i) \qquad (2.1)$$

is applied to the membership values and assigns realization $\vec{x}_i$ to the class k which attains highest confidence.

These membership values estimated by the various kinds of classification methods can be separated into two groups:

- *Membership probabilities* $p_{i,k} := P_{\text{method}}(C = k | \vec{X} = \vec{x}_i)$ which claim to cover the uncertainty in assessing that an observation $\vec{x}_i$ belongs to a particular class k, called *assessment uncertainty*. Regularly, statistical classification methods estimate such probabilities and output them in a probability matrix $\mathbf{P} \sim (N, K)$ with elements $p_{i,k} \in [0, 1]$. The rows of matrix $\mathbf{P}$ are probability vectors $\vec{p}_i := (p_{i,1}, \ldots, p_{i,K})'$ which sum for each observation $\vec{x}_i$ up to one.

- *Unnormalized scores* $s_{i,k} := s_{\text{method}}(C = k | \vec{X} = \vec{x}_i)$, usually given by Machine Learners in a score matrix $\mathbf{S} \sim (N, K)$. In contrast to probabilities, neither do the scores necessarily lie in the interval $[0, 1]$ nor sum the score vectors $\vec{s}_i = (s_{i,1}, \ldots, s_{i,K})'$ up to one.

Naturally, a classification method just outputs one type of membership value. An equivalent decision is achieved with the application of the classification rule to either membership probabilities or unnormalized scores, so that membership values m_{ik} in Equation (2.1) are replaced with $p_{i,k}$ and $s_{i,k}$, respectively. A motivation for preference of probabilities to unnormalized scores as membership values will be given in Section 2.2 followed by some requirements the membership probabilities have to meet, see Section 2.3. Detailed calibration methods for two–class tasks are introduced in Chapter 4 while methods for K classes follow in Chapter 5.

2.2. Post–processing with Membership values

A classifier is usually just part of a decision process in which decisions are associated with certain costs. If the classifier is involved in a cost–sensitive decision with costs differing between classes, it is desirable that the classifier generates membership values which cover the assessment uncertainty of an observation belonging to a particular class. Another advantage of a probabilistic membership value is that it simplifies the comparison and combination of results from different classifiers, see Duda *et al.* (1973).

2.2.1. Comparison and combination of multiple classifiers

In the days of data mining the number of competing classification methods is growing steadily. Naturally, there does not exist a universally best–performing classification rule and all classifiers have got their advantages and disadvantages. Hence, it is desirable to compare their performance. Therefore, it is not sufficient to regard only the precision of a classification method but also the quality of the membership values has to be considered in the comparison.

The best way to ensure comparable membership values is to generate membership probabilities, since probabilities are consistent in their attributes. A comparison of unnormalized scores instead would be a comparison of values with different attributes which would not make sense. Furthermore, an assessment uncertainty expressed in probabilistic terms is easy to understand and comprehensible for the user.

Additionally, if classifier outputs are comparable an accurate combination of classifiers can be realized. A combination of several different classifiers can lead to improved goodness, since different methods have differing strengths.

2.2.2. Cost–sensitive decisions

As stated before, the creation of classifier membership values is just one step in a decision process. Decisions based on the results of a classifier can be cost–sensitive, for example in clinical trials or in determination of credit risks. For such decisions it is helpful to obtain membership probabilities which can be combined with the referring costs because probabilities cover the uncertainty in the decision for a particular class. By using probabilities in post–processing it is guaranteed that costs are weighted consistently.

In cost–sensitive decisions it is not sufficient to use normalized membership values which simply meet the mathematical requirements for probabilities, i. e. sum up to one and lie in the interval $[0, 1]$. Such normalized membership values can have too extreme values, see e. g. Domingos & Pazzani (1996) and Zadrozny & Elkan (2001b), which would lead to an inappropriate weighting.

2.2.3. Multi–class problems

In Multi–class problems with number of classes $K > 2$ a regularization method usually reduces the classification problem to several binary tasks with either the one–against rest or the all–pairs approach, see Chapter 5 for details. These approaches generate for each observation several unnormalized scores per class which have to be combined to just one membership value. Since these scores are not normalized and do not reflect a probabilistic confidence they are not comparable and can give no basis for a reasonable combination, see Zhang (2004). Hence, these score vectors have to be calibrated to membership probability vectors before the combination to one matrix.

2.3. The aim of calibration

For the reasons described in Section 2.2, it is desirable that a classification method outputs membership values that reflect the assessment uncertainty of an observation belonging to a particular class. The solution for methods which do not generate probabilistic membership values is to transform or *calibrate* these unnormalized scores into the probability space. The aim of a calibration method is to scale the membership values into probabilities which are reliable measures for the uncertainty of assessment and give a realistic impression of the performance the rule yields on the test set. These probabilities are given in the matrix of calibrated probabilities

$$\hat{\mathbf{P}} := \begin{pmatrix} \hat{p}_{1,1} & \cdots & \hat{p}_{1,K} \\ \vdots & \ddots & \vdots \\ \hat{p}_{N,1} & \cdots & \hat{p}_{N,K} \end{pmatrix}$$

with elements $\hat{p}_{i,k} := \hat{P}_{\text{cal}}(C = k|\vec{m}_i)$ representing the probabilistic confidence in assignment of observation x_i to class k given a generated membership value $\vec{m}_i$. Regarding a calibrator as vector–valued function, the co–domain of a calibration function is the Unit Simplex $\mathcal{U} = [0, 1]^K$. This space, called *Standardized Partition Space* by Garczarek (2002), contains all combinations of real numbers that sum up to one and lie in the interval $[0, 1]$.

It can be visualized for three classes in a triplot barycentric coordinate system

showing how membership probabilities partition the observations according to
their class assignment, see Figure 2.1. Such a diagram is well known in the theory of experimental designs, see Anderson (1958), and was introduced to the
evaluation of membership probabilities by Garczarek & Weihs (2003).

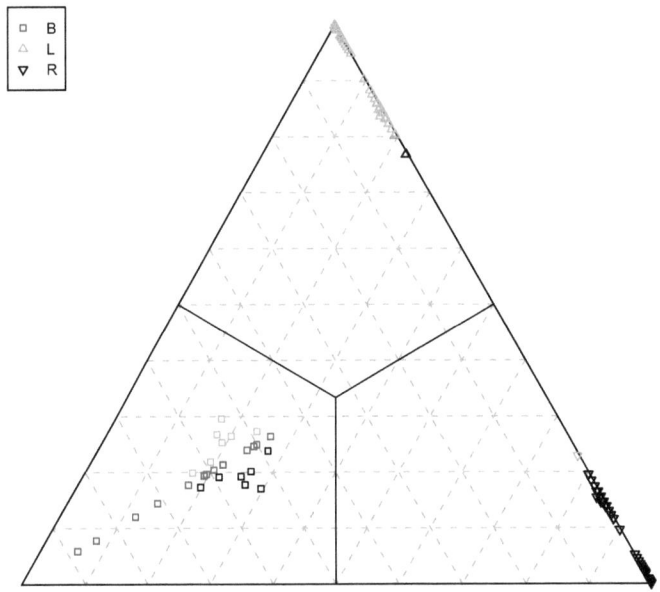

Figure 2.1.: Example for a barycentric plot for three classes

A barycentric plot illustrates the set of membership probability vectors $\vec{p}_1, \vec{p}_2, \vec{p}_3$
with probabilities between 0 and 1 and $p_{i,1}+p_{i,2}+p_{i,3} = 1$ by using a triangle. The
triangle is spanned by $(1,0,0)$, $(0,1,0)$ and $(0,0,1)$ in a three–dimensional space.

The three dimensions $\vec{p}_1, \vec{p}_2, \vec{p}_3$ correspond to the lower left, upper and lower right corner of the plot, respectively. The higher the confidence is in assignment to class 1, the closer the point is to the lower left corner. Points on the opposite triangle side, e. g. for class 1 the right side, have a membership probability of zero for this class. In the same way the other classes correspond to the respective triangle sides.

The grid lines show the observations at which one membership probability is constant, horizontal lines for example contain observations with an equal probability for class 2. Furthermore, the thick inner lines indicate the decision borders between two classes. Thus, these lines partition the triangle into three sectors in which each sector represents the assignment to the corresponding class.

While the type of symbol indicates the class which is assigned to, the color of the symbols indicate the true class of an object. Thus, symbols of one kind only occur in the respective sector while colors can occur in every sector. A mismatch of symbol, i. e. sector, and color indicates an object which is classified incorrectly. With the use of a stylized three–dimensional plot in shape of a tetrahedron, the visualization can be even extended for up to four dimensions and hence four classes, see Figure 2.2.

Here, the fourth class corresponds to the peak of the tetrahedron which has got the coordinates $(0,0,0,1)$ and lies in the middle of the plot. The idea and the interpretation is equivalent to the barycentric plot for three classes, see above.

2.4. Performance of classification rules

In learning a classification rule for a particular classification problem it is desirable to find the best rule, i. e. the rule $\hat{c}$ for which the L–risk $R_L(c,\hat{c})$, see Definition 2.1.6, is minimal for all classes $c \in \mathcal{C}$. Analogously, one can determine the optimal rule as the rule for which the counterpart of the risk, the expected precision, is maximal.

Since it is not possible to observe a whole population, such an optimal classification rule is learned on the basis of a finite training set. Certainly, the task in classification is to construct a rule which is not only well–performing for this training set but which can also be generalized for the classification of new data

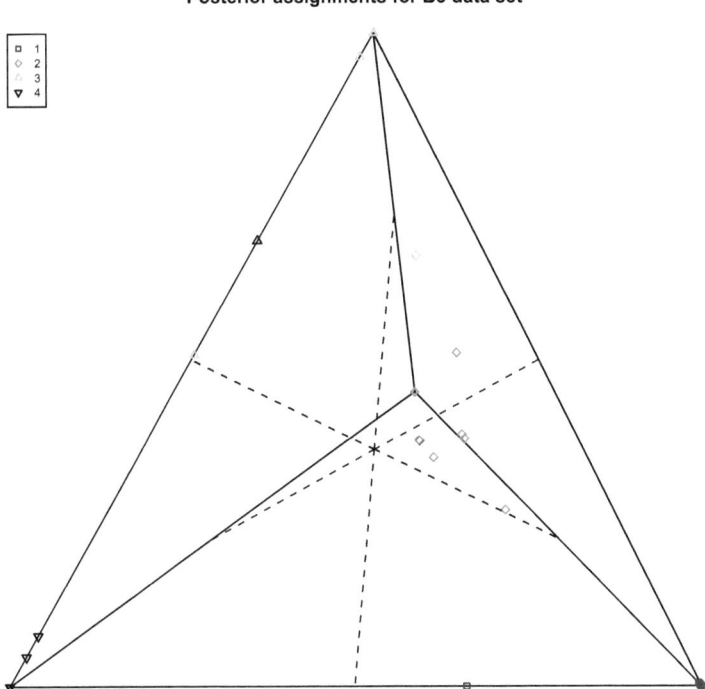

Figure 2.2.: Example for a barycentric plot for four classes

with unknown class labels. Thus, the classification rule on the one hand and the empirical correctness measure which approximates the expected precision on the other hand have to be determined by different observations to make performance measures reliable estimates for the generalization performance.

In Section 2.4.2 the three procedures simple validation, hold–out sets and cross–validation for calculating precision estimates for generalization will be introduced. The Theory presented here is based on Chapter 9 in Duda *et al.* (1973) and on Chapter 7 in Hastie *et al.* (2001). Section 2.4.3 illustrates how these methods can be additionally used for an appropriate learning of classification rules. Moreover, Section 2.4.4 introduces additive performance measures for a better evaluation of membership probabilities.

2.4.1. Goodness estimates

The aim of a classification method is usually to find a rule which has a minimal empirical L–risk

$$r_L(c, \hat{c}) := \frac{1}{N} \sum_{i=1}^{N} L[c_i, \hat{c}(\vec{x}_i)]. \qquad (2.2)$$

This empirical L–risk can be regarded as a stochastic approximation of the true L–risk $R_L(c, \hat{c})$, see Definition 2.1.6. Since the assignment of an observation to a class can be either correct or not, it is natural in classification problems to choose the 0–1–loss

$$L_{01}(c, \hat{c}) := \begin{cases} 0 & \hat{c} = c \\ 1 & \hat{c} \neq c \end{cases}$$

as loss function. Incorporating L_{01} in (2.2) leads to the empirical 0–1–risk r_{01}, called *empirical classification error*. r_{01} calculates the proportion of misclassified examples and is the standard measure for the imprecision of a classification rule. Since the counterpart of imprecision is of interest, one has to count the proportion of correctly assigned observations instead. With a training set $\mathcal{T} = \{(\vec{x}_i, c_i) : i = 1, \ldots, N\}$ as basis the empirical measure for the precision of a classification method is the *correctness rate*

$$\mathbf{CR} := 1 - r_{01} = \frac{1}{N} \sum_{i=1}^{N} \mathbf{I}_{[\hat{c}(\vec{x}_i) = c_i]}(\vec{x}_i). \qquad (2.3)$$

All sorts of different classification methods, see e. g. Hastie *et al.* (2001) or Hand (1997), can be compared with respect to this precision criterion.

Machine Learning methods like the Support Vector Machine and Artificial Neural Networks, see Sections 3.1 and 3.2, respectively, try to directly minimize the empirical L–risk (2.2) with some optimization algorithm. In this case it is not appropriate to incorporate the 0–1–loss L_{01}, since an optimization problem which minimizes the empirical classification error r_{01} is *Non–deterministic Polynomial–time hard (NP–hard)* and hence computationally intractable, see Kearns *et al.* (1987). Therefore, these classifiers usually minimize an L–risk based on a convex surrogate for L_{01}. The chosen loss function has got an effect on the probability information of the generated membership values. Therefore, Section 3.3 demonstrates the connection between the choice of loss function and the ability to estimate membership probabilities.

Besides, most of the classification methods can be regarded as rules which minimize a certain L-risk, at least for the dichotomous case, see Table 3.1 in Section 3.3.

2.4.2. Estimating performance reliably

The training set, see Definition 2.1.7, is the basis for the learning of a classification rule where usually some precision criterion like the correctness rate is maximized. Therefore, it is not advisable to evaluate the performance of the learned rule on the basis of the training set, since a correctness rate based on the training set observations is not an appropriate estimate for the generalization precision of a classifier. A straightforward approach to estimate performance reliably is the application of *simple validation*.

Definition 2.4.1 (Simple validation)
In *simple validation* a classification rule $\hat{c}$ is learned on the basis of the drawn training set $\mathcal{T} = \{(\vec{x}_i, c_i) : i = 1, \ldots, N\}$. Additionally, a further sample, called validation set, $\mathcal{V} = \{(\vec{x}_{i'}, c_{i'}) : i' = 1, \ldots, N_\mathcal{V}\}$ is drawn independently. The idea of simple validation is to apply the learned classification rule $\hat{c}$ to the validation set observations $\vec{x}_{i'}$ for estimation of their classes. The goodness of the classification rule is quantified by the correctness rate (2.3) on the basis of the true classes in the validation set.

As stated in Definition 2.4.1 it is necessary in simple validation to draw two samples independently. In real–life surveys this is often not possible and there is just one sample available. An option to supply two more or less independent data sets is to use so called *hold–out sets*. Here two thirds of the observations are drawn independently from the data set and are used as training set while the remaining third of the observations is used as validation set.

While the classification rule is learned on the basis of the training set, the validation set is ignored for the learning and is just used for the prediction of the correctness rate. Figure 2.3 illustrates how the behavior of the correctness rate differs with varying complexity of the fitted model between training and validation set, see also Weiss & Kulikowski (1991). This stylized function clarifies the necessity of using a validation set in predicting reliable correctness estimates.

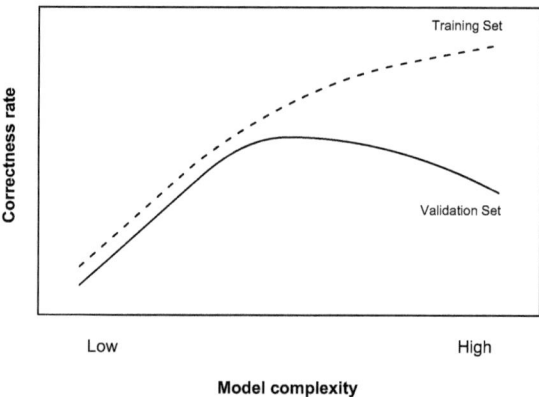

Figure 2.3.: **CR** for validation and training set as the model complexity is varied

More precise estimates can be derived with *cross–validation* which is a simple generalization of the hold–out method presented above.

Definition 2.4.2 (m–fold cross validation)
In *m–fold cross validation* the training set $\mathcal{T}$ is randomly divided into m disjoints of as nearly equal size as possible. The classifier is trained m times, each time with a different set held out as validation set, with the remaining $m - 1$ sets. Hence, each observation is part of a validation set once. The estimated correctness rate

is the average correctness determined by all observations.

Cross–validation is parsimonious with data, since every observation is used exactly once for validation. This parsimony leads to the fact that cross–validation is more precise than simple validation, especially for small data sets. Since 10–fold cross–validation is recommended for analyses, see e. g. Chapter 2 in Weiss & Kulikowski (1991) and Chapter 9.6 in Duda *et al.* (1973), it is used for the experiments in Chapter 6. Therefore, cross–validation is preferred to its major competitor Bootstrapping, see Chapter 7.11 in Hastie *et al.* (2001). Cross–validation can also be used in learning of classification rules, see Section 2.4.3, and in the determination of reliable estimates for the further performance measures introduced in Section 2.4.4.

2.4.3. Learning of rules and the problem of overfitting

Learning a classification rule just by optimizing the correctness rate (2.3) for the training set would lead to a classification rule which covers all features of the training set and is only optimal for this special data. Since the training set observations are just a small extract from the unknown true population with the same statistical properties, but not exactly the same values, such a learned rule would be far from optimal for the whole population, see Figure 2.3. This problem, called *overfitting*, leads to the fact that the correctness rate estimated on behalf of the training set is not generally valid, see the previous section. Overfitting is quite a drawback for the user, since the aim of learning classification rules is to find the best rule for the application on new data and not to find the best fit for some training data for which the true class labels are already available.

Since low generalization error is the ultimate goal for a classification method, overfitting is avoided in practice by learning the classifier until a maximum correctness rate or an optimal analogous criterion is attained for a validation set or in cross–validation. Basically, there exist two different methods which are applied in the learning of classification rules to avoid overfitting. This can either be a model which includes a complexity term into the objective, since overfitted models are too complex, see Figure 2.3, or the use of a *grid search*. A grid search is a simple method for the learning in which for each parameter several candidates are tested and the particular combination of parameters is chosen which attains the highest precision measure in the validation process. It is also possible to combine these two methods, see for example the regularization methods in Chapter 3.

2.4.4. Further Performance Measures

Naturally, the precision of a classification method, see Section 2.4.1, is the major characteristic of its performance. However, the goodness of a classification technique covers more than just correct classification.

To determine the quality of membership probabilities and hence the performance of a classification or calibration method in comparison to other membership probabilities several measures are necessary. Such a comparison of membership probabilities just on the basis of the correctness rate (2.3) means a loss of information and would not include all requirements a probabilistic classifier score has to fulfill. To overcome this problem, calibrated probabilities should satisfy the two additional axioms:

- Effectiveness in the assignment and

- Well–Calibration in the sense of DeGroot & Fienberg (1983).

In the following these two axioms will be elaborated in detail.

Effective assignment Membership probabilities should be effective in their assignment, i. e. moderately high for true classes and small for false classes. An indicator for such an effectiveness is the complement of the *Root Mean Squared Error*:

$$1 - \mathbf{RMSE} := 1 - \frac{1}{N}\sum_{i=1}^{N}\sqrt{\frac{1}{K}\sum_{k=1}^{K}\left[\mathbf{I}_{[c_i=k]}\left(\vec{x}_i\right) - P\left(c_i = k | \vec{x}\right)\right]^2}. \quad (2.4)$$

The **RMSE** is equivalent to other performance measures, such as the *Brier Score* introduced by Brier (1950) or the *Accuracy* introduced by Garczarek (2002). These measures also base on the squared differences between membership probabilities and an indicator function for the true class.

Well–Calibrated probabilities DeGroot & Fienberg (1983) give the following definition of a well–calibrated forecast: "If we forecast an event with probability p, it should occur with a relative frequency of about p." To transfer this requirement from forecasting to classification it is required to partition the training/test set according to the class assignment into K groups $\mathcal{T}_k := \{(c_i, \vec{x}_i) \in \mathcal{T} : \hat{c}(\vec{x}_i) = k\}$ with $N_{\mathcal{T}_k} := |\mathcal{T}_k|$ observations. Thus, in a partition $\mathcal{T}_k$ the forecast is class k. Considering Figure 2.1, these partitions are equivalent to the sectors in the barycentric plot.

Predicted classes can differ from true classes and the remaining classes $j \neq k$ can actually occur in a partition $\mathcal{T}_k$. To cover these individual misclassifications the average confidence $\mathbf{Cf}_{k,j} := \frac{1}{N_{\mathcal{T}_k}} \sum_{x_i \in \mathcal{T}_k} P(k|\hat{c}(\vec{x}_i) = j)$ is estimated for every class j in a partition $\mathcal{T}_k$. According to DeGroot & Fienberg (1983) this confidence should converge to the average correctness $\mathbf{CR}_{k,j} := \frac{1}{N_{\mathcal{T}_k}} \sum_{x_i \in \mathcal{T}_k} \mathbf{I}_{[c(\vec{x}_i)=j]}$. The average closeness of these two measures

$$\mathbf{WCR} := 1 - \frac{1}{K^2} \sum_{k=1}^{K} \sum_{j=1}^{K} |\mathbf{Cf}_{k,j} - \mathbf{CR}_{k,j}| \qquad (2.5)$$

indicates how well–calibrated the membership probabilities are. Hence, this measure will be called *Well–Calibration Ratio* in the following.

Calibration measure On the one hand, the minimizing "probabilities" for the **RMSE** (2.4) can be just the class indicators especially if overfitting occurs in the training set. On the other hand, the **WCR** (2.5) is maximized by vectors in which all membership probabilities are equal to the individual correctness values. Based on the idea of desirability indices by Harrington (1965), it is convenient to overcome these drawbacks with combining the two calibration measures by their geometric mean. This yields the calibration measure

$$\mathbf{Cal} := \sqrt{(1 - \mathbf{RMSE}) \cdot \mathbf{WCR}} \qquad (2.6)$$

which indicates how well the membership probabilities reflect the assessment uncertainty.

Ability to separate Finally, another performance measure is introduced by Garczarek (2002). In the following it is explained why this measure is not regarded in the experiments of Chapter 6.

The *ability to separate* measures how well classes are distinguished by the classification rule. The measure is based on the non–resemblance of classes which is the counterpart of the concept of resemblance by Hand (1997).

Classes are well distinguished by a classification rule if an membership probability approaches 1 for the assigned class and is close to 0 for the classes which the rule not assigns to. In contrast to the **RMSE** (2.4), the ability to separate is based on the distance between a membership probability for a particular class and an Indicator function indicating whether this class is the assigned one. Analogous to the **RMSE**, the squared distances are summed for every class over all observations. The sum is standardized so that a measure of 1 is achieved if all observations are assigned without insecurity:

$$\mathbf{AS} := 1 - \frac{K}{K-1} \frac{1}{N} \sum_{i=1}^{N} \sqrt{\sum_{k=1}^{K} \left[\mathbf{I}_{[\hat{c}_i = k]}(\vec{x}_i) - P(C = k | \vec{x}) \right]^2}. \quad (2.7)$$

A high ability to separate implies that the classification rule works out the characteristic differences between the classes, but not necessarily.

The ability to separate has got the major drawback that it does not consider the true classes which can lead to misinterpretations. A high **AS** does not necessarily mean that the membership probabilities are of high quality, because it just means that the probabilities for the assigned class are high and the other ones are low. If the ability to separate is high and $1 - $ **RMSE** and/or correctness rate are small it is an indicator for the fact that the probabilities are too extreme and do not reflect the assessment uncertainty. Thus, this measure is omitted in the experimental analysis in Chapter 6.

3. Regularization methods for classification

This chapter gives an overview of the two standard *Regularization Methods* for classification, Support Vector Machine and Artificial Neural Networks, in Section 3.1 and 3.2, respectively. These methods have in common that a classification rule (2.1) is learned by minimizing a regularized empirical risk (2.2) based on a convex loss function.

Subsequently, it is shown in Section 3.3 which kind of loss function has to be chosen in regularization so that unnormalized scores contain sufficient probability information for a calibration into membership probabilities. Since regularization methods apply loss functions which are convex surrogates of the 0–1–Loss, these methods are initially constructed for two–class tasks. While the optimization algorithm for an Artificial Neural Network has been extended for K–class tasks, see Section 3.2, the idea underlying the Support Vector Machine with maximizing the margin between two classes makes a reduction to binary classification problems necessary, see e. g. Vogtländer & Weihs (2000) as well as Duan & Keerthi (2005) for an overview. All so–called Multivariate Support Vector Machines as introduced in Allwein *et al.* (2000), Crammer (2000), Dietterich & Bakiri (1995), Lee *et al.* (2004) or Weston & Watkins (1998) base on a reduction to binary classification problems.

Chapter 4 shows the several existing methods for calibrating two–class membership values into probabilities that reflect the assessment uncertainty while generalization algorithms for K classes are presented in Chapter 5.

3.1. Support Vector Machine

The *Support Vector Machine (SVM)*, introduced by Cortes & Vapnik (1995), is a Machine Learning method which can be used for regression as well as for dichotomous classification tasks. The Support Vector Classifier separates the two classes by an optimal hyperplane which maximizes the distance between the observations for either class, see Section 3.1.1. If classes are not perfectly separable, the optimal hyperplane is constructed so that the amount of misclassification is hold as small as possible. Two major approaches have been introduced in recent years, $L1$– and $L2$–SVM, which basically differ in the penalization amount of wrongly classified observations.

To enable for both approaches the possibility of creating a discriminant which is not necessarily linear, attribute variables are mapped into a higher dimensional feature space for separation, see Section 3.1.2.

The optimization procedure which bases for the $L2$–SVM on a grid search and cross–validation is presented in Section 3.1.3.

Since the SVM procedure is only directly applicable for two classes, a multi–class problem with number of classes $K > 2$ is reduced to several binary tasks, see Chapter 5. The SVM theory presented in this section is mainly based on Vapnik (2000), Burges (1998) and Cristianini & Shawe-Taylor (2000).

3.1.1. Separating Hyperplanes for two classes

This section shows how the SVM constructs an optimal separating hyperplane for two perfectly separated classes as well as the generalization for the non–separable case.

A hyperplane H is generally defined by

$$H = \{\vec{x} : f(\vec{x}) = \vec{w}' \cdot \vec{x} + b = 0\}$$

where $\vec{w}$ is a unit vector, i. e. a vector of length $\|\vec{w}\| = 1$.

Given a training set $\mathcal{T} = \{(\vec{x}_i, c_i) : i = 1, \ldots, N\}$ with binary class variable $c_i \in \mathcal{C} := \{-1, +1\}$ it is the aim to find a hyperplane, i. e. a decision function f, as basis for the classification rule

$$\hat{c}(\vec{x}) \; := \; \text{sign}[f(\vec{x})] = \text{sign}[\vec{w}' \cdot \vec{x} + b] \, . \tag{3.1}$$

The decision function $f(\vec{x})$ can be used to determine a rule which is analog to the general classification rule (2.1) and assigns an observation $\vec{x}$ to the class with highest membership value. Therefore, a membership value for the positive class becomes $m(+1|\vec{x}) := f(\vec{x})$ and hence the membership value for the negative class is the negative term $m(-1|\vec{x}) := -f(\vec{x})$.

Perfectly separable classes While Linear Discriminant Analysis, see Fisher (1936), constructs a linear discriminant on the basis of the assumption of multivariate normality, the SVM separates the two classes by an optimal hyperplane H which maximizes the *margin d* from the hyperplane to the closest points from either class instead. These points lie on the two additional parallel hyperplanes $H_+ := \{\vec{x} : f(\vec{x}) = +1\}$ and $H_- := \{\vec{x} : f(\vec{x}) = -1\}$, respectively. Thus, H_+ and H_- are parallel to H with margin d, see Figure 3.1.

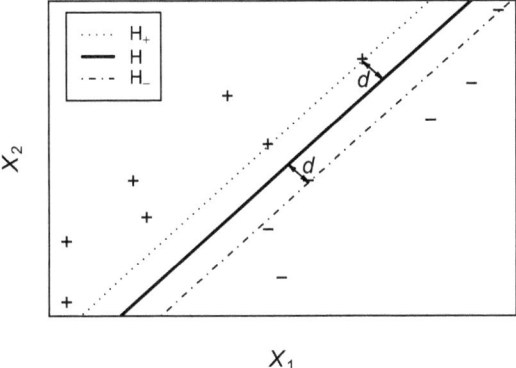

Figure 3.1.: Example of an optimal hyperplane for the perfectly separable case

If classes are perfectly separable, an optimal decision function f can be found so that every observation is assigned correctly, i. e. $c_i f(\vec{x}_i) \geq 0 \ \forall \ i = 1, \ldots, N$. Hence, it is possible to find a hyperplane H which creates the biggest margin d to the training points for the positive and the negative class, respectively, see

Figure 3.1. This search can be formulated in the optimization problem

$$\max_{\substack{\vec{w},b \\ \|\vec{w}\|=1}} d \qquad (3.2)$$

with subject to $c_i\left(\vec{w}'\vec{x}_i + b\right) \geq d, i = 1,\ldots,N$. Hastie et al. (2001) show that this optimization problem can be rephrased by arbitrarily setting the norm of the *normal vector* $\|\vec{w}\| = 1/d$ into the minimization of the functional

$$\Phi(\vec{w}) \;=\; \|\vec{w}\| \qquad (3.3)$$

subject to the constraint

$$c_i\left(\vec{w}'\vec{x}_i + b\right) \geq 1, \quad i = 1,\ldots,N\,. \qquad (3.4)$$

In the following, the optimization procedure will be only shown for the non–separable case, see below, since this is a generalization of the perfectly separable case.

Non–separable classes In real–life data sets classes usually overlap in the attribute space so that they are not perfectly separable. If this is the case, misclassifications have to be possible in constructing the optimal separating hyperplane H. A hyperplane of this type is called *soft–margin hyperplane*, see Cortes & Vapnik (1995). The SVM approach to create a soft–margin hyperplane is to modify the functional (3.3) by incorporating an error term

$$\Phi(\vec{w}) \;=\; \frac{1}{2}\|\vec{w}\|^2 + \frac{\gamma}{2}\sum_{i=1}^{N} L(c_i, \vec{w}'\vec{x}_i + b) \qquad (3.5)$$

with some loss function $L : \mathbb{R} \to [0,\infty[$ and a user–defined regularization parameter γ, see Burges (1998). A large value for γ corresponds to a high penalty for errors. An error penalized by the loss function is induced by an observation which lies on the wrong side of its corresponding hyperplane H_- or H_+.

A natural choice for the loss function would be the 0–1–loss L_{01} (2.3), but optimization problems based on non–convex loss functions like L_{01} are computationally intractable, see Kearns et al. (1987). To simplify the computation problem Cortes & Vapnik (1995) apply the *q–hinge loss* function

$$L_{q\text{–hinge}}[c_i, f(\vec{x}_i)] \;:=\; \max\{0, 1 - c_i f(\vec{x}_i)\}^q \qquad (3.6)$$

with positive integer q is used as convex surrogate loss function. Including the q–hinge loss into the optimization problem (3.5) leads to the functional:

$$\Phi(\vec{w},\vec{\xi}) = \frac{1}{2}\|\vec{w}\|^2 + \frac{\gamma}{2}\sum_{i=1}^{N}\xi_i^q \qquad (3.7)$$

where the vector of non–negative *slack variables* $\vec{\xi} = (\xi_1,\ldots,\xi_n)'$ is the output generated by the maximum term in the loss function (3.6). A slack variable ξ_i measures the proportional amount for which observation $\vec{x}_i$ is on the wrong side of the corresponding hyperplane H_- or H_+, see Figure 3.2.

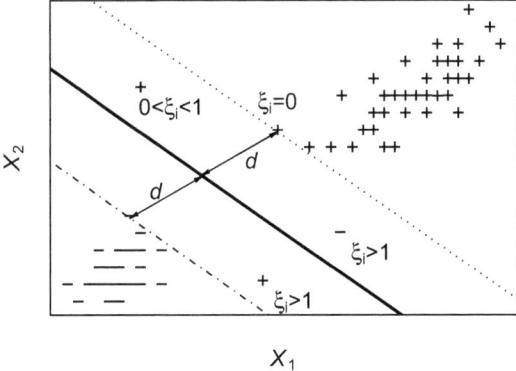

Figure 3.2.: Example for SVM with slack variables for non–separable case

Therefore, the constraint applicable in perfect separation (3.4) can not be met for all observations and has to be modified into

$$c_i\left(\vec{w}'\vec{x}_i + b\right) \geq 1 - \xi_i, \quad i = 1,\ldots,N \qquad (3.8)$$

by using the slack variables which have to hold the condition

$$\sum_{i=1}^{N}\xi_i \leq \text{constant}. \qquad (3.9)$$

This leads to the fact that for $\xi_i < 1$ the corresponding observation $\vec{x}_i$ is still classified correctly, while $\xi_i > 1$ implies misclassification, see Figure 3.2. With bounding the sum of slack variables (3.9) the total proportional amount of observations on the wrong side of either margin and the total number of misclassified observations are bound as well.

In specifying the parameter q of the q–hinge loss function in (3.6) it is common to set $q = 1$, see Cortes & Vapnik (1995). The corresponding classifier is called *L1–SVM*. In recent years, the competitor *L2–SVM* introduced by Suykens & Vandewalle (1999) which is based on the *quadratic hinge loss*, i. e. setting $q = 2$, has gained interest because of some better statistical properties, see Section 3.3. Consequently, only the optimization procedures for *L2–SVM* will be shown in the following.

L2–SVM To find the solution for the optimization of the functional (3.7) with setting $q = 2$ the constraint (3.8) is incorporated into the *primal Lagrange functional*

$$\mathcal{L}_p = \frac{1}{2}\|\vec{w}\|^2 + \frac{\gamma}{2}\sum_{i=1}^{N}\xi_i^2 - \sum_{i=1}^{N}\alpha_i\left[c_i(\vec{w}'\vec{x}_i + b) - (1 - \xi_i)\right] \quad (3.10)$$

where the α_i are *Lagrangian multipliers*, see Cristianini & Shawe-Taylor (2000). Compared to *L1–SVM*, see Burges (1998), a positivity constraint for the slack variables ξ_i is not necessary, since slack variables are squared in the functional, see Shawe-Taylor & Cristianini (2004).

$\mathcal{L}_p$ has to be minimized with respect to $\vec{w}$, b and ξ_i. Hence, the respective derivatives

$$\frac{\partial \mathcal{L}_p}{\partial \vec{w}} = \vec{w} - \sum_{i=1}^{N}\alpha_i c_i \vec{x}_i \stackrel{!}{=} 0 \quad (3.11)$$

$$\frac{\partial \mathcal{L}_p}{\partial b} = -\sum_{i=1}^{N}\alpha_i c_i \stackrel{!}{=} 0 \quad (3.12)$$

$$\frac{\partial \mathcal{L}_p}{\partial \xi_i} = \gamma\xi_i - \alpha_i \stackrel{!}{=} 0 \quad (3.13)$$

are set equal to zero. Incorporating the solutions of the derivative equations (3.11) – (3.13) in the primal Lagrangian $\mathcal{L}_p$ (3.10) constitutes the *Lagrange Wolfe Dual*

$$\begin{aligned}\mathcal{L}_D &= \sum_{i=1}^{N}\alpha_i - \frac{1}{2}\sum_{i=1}^{N}\sum_{j=1}^{N}\alpha_i\alpha_j c_i c_j \vec{x}_i'\vec{x}_j - \frac{1}{2\gamma}\sum_{i=1}^{N}\alpha_i^2 \\ &= \sum_{i=1}^{N}\alpha_i - \frac{1}{2}\sum_{i=1}^{N}\sum_{j=1}^{N}\alpha_i\alpha_j c_i c_j \left(\vec{x}_i'\vec{x}_j + \frac{1}{\gamma}\delta_{ij}\right)\end{aligned} \quad (3.14)$$

with Kronecker–delta defined as $\delta_{ij} := 1$ if $i = j$ and 0 else.

According to Kuhn & Tucker (1951) the minimum of the primal Lagrangian (3.10) is given by the maximum of the Lagrange Wolfe Dual under constraints (3.8) and

$$\sum_{i=1}^{N} \alpha_i c_i = 0, \quad (3.15)$$

$$\alpha_i \geq 0, \quad (3.16)$$

$$\alpha_i \left[c_i \left(\vec{w}' \vec{x}_i + b \right) - (1 - \xi_i) \right] = 0. \quad (3.17)$$

Condition (3.15) is derived by solving (3.12) while condition (3.16) follows from the solution of the derivative for the slack variables (3.13). Finally, (3.17) is the complementary *Karush–Kuhn–Tucker condition* based on the Theorem by Kuhn & Tucker (1951).

By solving the derivative for $\vec{w}$ (3.11) one can obtain an equation for calculating the normal vector

$$\vec{w} = \sum_{i=1}^{N} \alpha_i c_i \vec{x}_i \quad (3.18)$$

which is a linear combination of the training set observations for which $\alpha_i > 0$, namely the *support vectors*. Thus, the optimal hyperplane for separation is induced by

$$f(\vec{x}) = \sum_{i=1}^{N} \alpha_i c_i \vec{x}_i' \vec{x} + b. \quad (3.19)$$

The training set observations influence the learning of the optimal hyperplane just by the scalar product $\vec{x}_i' \vec{x}$, see Equations (3.14) and (3.19), which simplifies the generalization in Section 3.1.2.

3.1.2. Generalization into high–dimensional space

The previously presented optimization problem constructs an optimal hyperplane which separates two classes linearly. Since classes are not regularly separable by a linear discriminant, it is desirable to have a more flexible discriminant which is not necessarily linear. For the SVM Schölkopf *et al.* (1995) introduced the idea of mapping the attribute variables $\vec{x}$ from the *attribute space* $\mathcal{X}$ into a higher dimensional Hilbert space $\mathcal{H}$, called *feature space*. In this space an optimal separating hyperplane is constructed as described in Section 3.1.1.

According to the previous section a learned optimal hyperplane just depends on the training set observations in terms of the scalar product $\vec{x}_i'\vec{x}$. Hence, for constructing the optimal hyperplane in the feature space $\mathcal{H}$ it is only required to calculate the inner products between support vectors and vectors of the feature space. In a Hilbert space $\mathcal{H}$ the inner product

$$(\vec{z}_i \cdot \vec{z}) = K(\vec{x}, \vec{x}_i) \qquad (3.20)$$

with $\vec{z} \in \mathcal{H}$ and $\vec{x} \in \mathcal{X}$ can be regarded as a Kernel function K. According to Courant & Hilbert (1953) $K(\vec{x}, \vec{x}_i)$ has to be a symmetric positive (semi–) definite function.

Regular choices for K are

- d'th degree polynomial $K(\vec{x}, \vec{x}_i | d) = (1 + \langle \vec{x}, \vec{x}_i \rangle)^d$,
- Gaussian radial basis $K(\vec{x}, \vec{x}_i | \sigma) = \exp\left(-\dfrac{\|\vec{x} - \vec{x}_i\|^2}{2\sigma^2}\right)$,
- Neural Network $K(\vec{x}, \vec{x}_i | \kappa_1, \kappa_2) = \tanh\left(\kappa_1 + \kappa_2 \langle \vec{x}, \vec{x}_i \rangle\right)$,

see Hastie *et al.* (2001). In the following the Gaussian radial basis kernel is chosen, since it is widely recommended, see Hsu *et al.* (2003) and Garczarek (2002). According to Hsu *et al.* (2003) the radial basis kernel has less numerical difficulties than the polynomial kernel while the Neural Network kernel has the drawback that it is sometimes not valid, see Vapnik (2000). Additionally, it is an advantage that there is just one parameter to optimize for this Kernel function and the inclusion of a further parameter as in the Neural Network kernel might lead to a too complex model. Furthermore, according to Schölkopf *et al.* (1995) the choice of the Kernel function is not crucial but estimating the appropriate parameter σ is.

3.1.3. Optimization of the SVM Model

Summing the two previous sections up, the optimal separating hyperplane and corresponding optimal Lagrangian multipliers $\hat{\alpha}_i$ found by the $L2$–SVM method is the function which maximizes the Lagrange Wolfe Dual (3.14) under the constraints (3.8) and (3.15) – (3.17). The solution can be represented as the decision function

$$f(\vec{x}) = \sum_{i=1}^{N} \alpha_i c_i K(\vec{x}, \vec{x}_i | \sigma) + b.$$

To construct an hyperplane of this type which separates observations optimally, it is required to choose optimal parameters $\hat{\sigma}$ and $\hat{\gamma}$ in the optimization process. Estimates can be found by using a grid search and cross–validation, see below. Finally, it has been shown for example by Wahba (1998) and Schölkopf et al. (1997) that the SVM optimization procedure can be cast as a regularization problem in a *Reproducing Kernel Hilbert Space* (RKHS), see Aronszajn (1950). Thus, the SVM procedure using the Kernel function K can be seen as minimization of the empirical regularized q–hinge–risk

$$\begin{aligned} r_{q\text{-hinge}}^{reg} &:= \lambda \|f\|_H^2 + r_{q\text{-hinge}}[c_i, f(\vec{x}_i)] \\ &= \lambda \|f\|_{\mathcal{H}_K}^2 + \frac{1}{N} \sum_{i=1}^{N} L_{q\text{-hinge}}[c_i, f(\vec{x}_i)] \end{aligned} \quad (3.21)$$

with convex loss function $L_{q\text{-hinge}}$ (3.6) and $\mathcal{H}_K$ the RKHS with corresponding Kernel K. According to Hastie et al. (2001) the same solution for a function $f(\vec{x}) = h(\vec{x}) + b$ with $h \in \mathcal{H}_K$ and offset term $b \in \mathbb{R}$ is given by setting the regularization parameter $\lambda = 1/2\gamma$.

For optimizing the Lagrangian multipliers α_i in L2–SVM the conjugate gradient method is used, see Suykens et al. (1999). Furthermore, it is recommended by Hsu et al. (2003) to apply a loose grid search on parameters σ and γ with range $\gamma = 2^{-5}, 2^{-3}, \ldots, 2^{15}$ and $\sigma = \sqrt{1/2^{-15}}, \sqrt{1/2^{-13}}, \ldots, \sqrt{1/2^3}$ to find the best pair (σ, γ). Next step is to conduct a finer grid search in the area around the optimal pair of the loose grid search. Finally, the SVM is trained for the optimal pair of σ and γ with the presented algorithm.

3.2. Artificial Neural Networks

The theory of *Artificial Neural Networks (ANN)* emerged separately from three different sources:

- a biological interest of understanding the human brain;
- broader issues in artificial intelligence to copy human abilities like speech and use of language;
- a statistical approach to pattern recognition and classification.

Biological research shows that a neuron in the human brain behaves like a switch by activating a signal when sufficient neurotransmitter is accumulated in the cell body. These signals can travel parallel and serially through the brain along synaptic connections in the nervous system.

The fundamental idea of an Artificial Neural Network for classification is to copy these processes in the human brain. An ANN consists of a group of *nodes* which simulate neurons and are "connected" by a group of weights, analogous to the synaptic connections. Therefore, the idea of an ANN is to apply a linear function to the *input nodes* i. e., attribute variables $X_1, \ldots, X_p$, and regard the outcome as derived features. Based on the signal activation in the human brain the target variable $\vec{Y}$ is modeled as a non–linear threshold function g, called *activation function*, of the derived features. In a classification task with K classes the target variable $\vec{Y}$ is a vector of size K where each element is an indicator function for the corresponding class $k \in [1, \ldots, K]$:

$$Y_k := \mathbf{I}_{[c(\vec{X})=k]}(\vec{X}).$$

For classification tasks with $K > 2$ this procedure, also visualized in Figure 3.3, reduces the problem into K binary decisions for either class and is thus equivalent to the multivariate reduction method one–against rest, see Chapter 5.

The original model by McCulloch & Pitts (1943) and the further extension by Rosenblatt (1962) are presented in Section 3.2.1 and 3.2.2, respectively. The Learning rule is shown in Section 3.2.3 and the parameter optimization in Section 3.2.4.

3.2.1. McCulloch–Pitts neuron and Rosenblatt perceptron

The original model for an ANN by McCulloch & Pitts (1943) is quite similar to Discriminant Analysis by Fisher (1936), since the *McCulloch–Pitts neuron* derives features where the linear function is a weighted sum of input variable realizations $x_1, \ldots, x_p$. The activation function g_k which is applied to the derived features is a threshold function. This yields the model

$$y_k := g_k\left(\sum_{l=1}^{p} \alpha_{k,l} x_l\right) = \begin{cases} 1 & \text{if } \sum_{l=1}^{p} \alpha_{k,l} x_l - \alpha_{k,0} \geq 0 \\ 0 & \text{else} \end{cases}, \quad (3.22)$$

with threshold $\alpha_{k,0}$ and a vector of individual weights $\vec{\alpha}_k := (\alpha_{k,1}, \ldots, \alpha_{k,p})'$. For a schematic of the McCulloch–Pitts neuron see Figure 3.3.

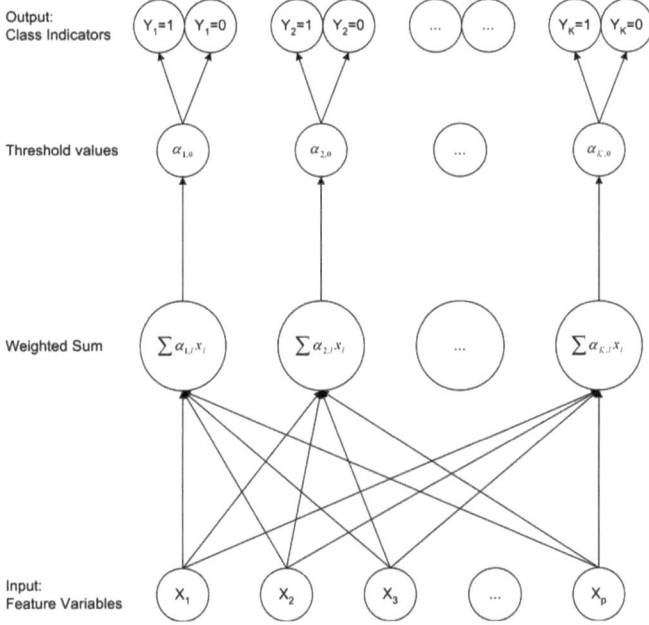

Figure 3.3.: McCulloch–Pitts neuron

Since the activation function g_k discontinuously jumps from a lower to an upper limiting value, this model (3.22) defines a non–linear function across a hyperplane in the attribute space. With such a threshold activation function the neuron output is 1 on one side of the hyperplane and 0 on the other one.

Rosenblatt (1958) introduced the *perceptron* structure in which the neurons segment the attribute space not only into two but into more regions. Here, the activation function of the McCulloch–Pitts neuron (3.22) is generalized to

$$y_k := g_k \left(\sum_{l=0}^{p} \alpha_{k,l} x_{l+1} \right) = g_k \left(\vec{\alpha}_k' \vec{x} \right) \qquad (3.23)$$

where the activation function g_k can be any non–linear function. The threshold values $\alpha_{k,0}$ of the McCulloch–Pitts neuron (3.22) are accommodated by including a vector of length N with constant term 1 as first column vector of observation matrix $\mathbf{X}$ with index 0, so that realization vectors become $\vec{x}_i := (1, x_{i,1}, \ldots, x_{i,p})'$. Estimated target values $\hat{y}_{i,k} := g_k \left(\vec{\alpha}_k' \vec{x}_i \right)$ generated for an individual observation $\vec{x}_i$ can be regarded as membership values $m(C = k | \vec{x}_i)$. Therefore, the general classification rule (2.1) is applicable for a target value $\hat{y}_{i,k}$ estimated by an ANN similarly to membership values generated by statistical classification methods. A particular observation $\vec{x}_i$ is assigned to the class with highest target value $\hat{y}_{i,k}$.

3.2.2. Multi Layer Perceptron

Rosenblatt (1962) proposed the *Perceptron Learning Rule* for learning suitable weights $\alpha_{k,l}$ for classification tasks. However, according to the analyses by Minsky & Papert (1969) the Rosenblatt–Perceptron does not cover many real world problems, since such networks are only capable for linearly separable classes. For example an *exclusive–or (XOR)* function on the input variables could not be incorporated by the Rosenblatt–Perceptron. Hence, Minsky & Papert (1969) suggested with the *Multi Layer Perceptron (MLP)* a further extension which is widely used today. The MLP is a two–stage classification model in which a first neuron maps from an *input layer* to a *hidden layer* and a second neuron maps from the hidden layer to an *output layer*, see Figure 3.4.

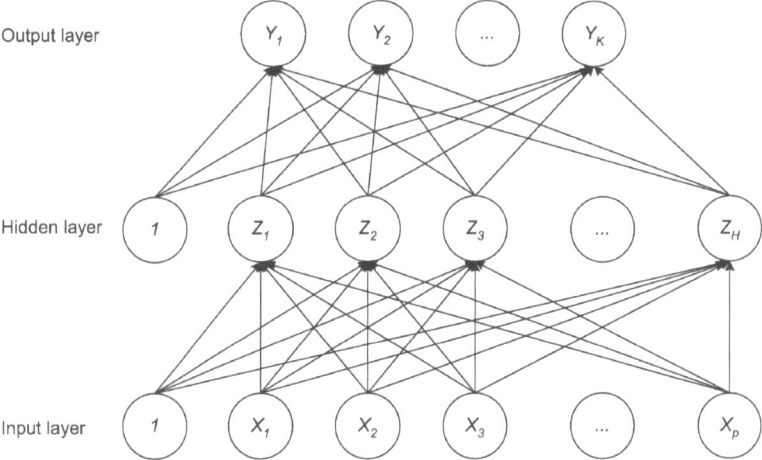

Figure 3.4.: Structure of the Multi Layer Perceptron

The starting point of the first stage of the MLP is the input layer which is formed by the observation matrix **X** including the additional column of 1's. A neuron as in (3.23) is used to derive the nodes in the hidden layer

$$z_h := g_h\left(\vec{\alpha}_h' \vec{x}\right), h = 1, \ldots, H \tag{3.24}$$

with vector of weights $\vec{\alpha}_h := (\alpha_{h,0}, \ldots, \alpha_{h,p})'$. These nodes are called *hidden units* since they are not directly observed. In contrast to the fixed number of nodes in the input and output layer, $p+1$ and K, respectively, the number H of hidden units is flexible. The optimal number of hidden units for a particular classification task can be found by a grid search, see Section 3.2.4.

The second stage of the MLP is a further neuron pointing from the vector of hidden units $\vec{z} := (1, z_1, \ldots, z_H)'$, including an obligatory added $z_0 = 1$ as in the input layer, to the target values

$$y_k := g_k^*\left(\vec{\beta}_k' \vec{z}\right), k = 1, \ldots, K \tag{3.25}$$

with vector of weights $\vec{\beta}_k := (\beta_{k,0}, \ldots, \beta_{k,H})'$.

Instead of choosing activation functions g and g^* as threshold functions like in the McCulloch–Pitts neuron (3.22) they are chosen as sigmoidal functions which satisfy the following conditions:

$$\begin{aligned} g(x) &\to 0 & \text{if } x \to -\infty \\ g(x) &\to 1 & \text{if } x \to \infty \\ g(x) &+ g(-x) = 1 & \end{aligned} \qquad (3.26)$$

Typically g and g^* are set to the logistic activation function

$$g(x) \;=\; \frac{1}{1+\exp(-x)} \;=\; \frac{\exp(x)}{1+\exp(x)} \qquad (3.27)$$

which meets condition (3.26).

There also exist MLP with more than one hidden layer and the optimal number of hidden layers depends on the particular classification task, but according to Cybenko (1989) every bounded continuous function can be approximated by an architecture with one sufficiently large hidden layer. Thus, a MLP with only one hidden layer should lead to appropriate results for most of the existing classification problems.

3.2.3. Generalized Delta Rule

Rumelhart et al. (1986) proposed the *Generalized Delta Rule* for estimation of weights in an ANN with MLP–structure. Therefore, all MLP–stages i. e., neurons of the ANN can be theoretically combined into a global decision function f_k of the observations. Hence, the model regards target values

$$\begin{aligned} y_k &:= f_k(\vec{x}, \vec{w}) \\ &= g_k^*\left[\beta_{k,0} + \sum_{h=1}^{H} \beta_{k,h} g_h\left(\vec{\alpha}_h' \vec{x}\right)\right] \end{aligned}$$

as a function of observation $\vec{x}$ and a set of complete set of unknown weights

$$\vec{w} \;:=\; \left\{\vec{\alpha}_h, \vec{\beta}_k : k = 1, \ldots, K; h = 1, \ldots, H\right\} .$$

The weights of an ANN for classification are estimated by minimizing an objective which bases on the L–risk

$$r_L\left(\vec{y}; \vec{f}\right) \;:=\; \frac{1}{N}\sum_{i=1}^{N} L\left[\vec{y}_i; \vec{f}(\vec{x}_i)\right]$$

with the general vector–valued function $\vec{f} := (f_1, \ldots, f_K)$.
A regular choice for the loss function in an ANN is the quadratic loss

$$\begin{aligned} L_{\text{quad}}\left(\vec{y}, \vec{f}\right) &:= \left\|\vec{y} - \vec{f}\right\|_2 \\ &= \sum_{k=1}^{K}(y_k - f_k)^2 \end{aligned}$$

where the corresponding risk is the *mean squared error*

$$r_{\text{quad}}\left(\vec{y}; \vec{f}\right) := \frac{1}{N}\sum_{i=1}^{N}\sum_{k=1}^{K}\left[y_{k,i} - f_k\left(\vec{x}_i\right)\right]^2 . \tag{3.28}$$

For convenience a proportional term, the *squared error*

$$SE\left(\vec{w}\right) = \sum_{i=1}^{N}\sum_{k=1}^{K}\left[y_{k,i} - f_k\left(\vec{x}_i, \vec{w}\right)\right]^2 \propto r_{\text{quad}}\left(\vec{y}; f_1, \ldots, f_K\right) \tag{3.29}$$

is used as objective instead.

The generic approach to minimize the squared error (3.29) is by a iteration procedure based on gradient descent. In this context this procedure is called *back–propagation*, see Rumelhart *et al.* (1986).

3.2.4. Regularization and grid search

As mentioned in Section 2.4.3 learning a classification rule just by optimizing a precision criterion for the training set leads to overfitting. Therefore, the squared error (3.29) is regularized with the additional term

$$J(\vec{w}) := \sum_{k=1}^{K}\sum_{h=0}^{H}\beta_{k,h}^2 + \sum_{h=1}^{H}\sum_{l=0}^{P}\alpha_{h,l}^2$$

to avoid overfitting and therefore the objective which has to be minimized in an ANN is

$$SE(\vec{w}) + \lambda J(\vec{w}) . \tag{3.30}$$

The optimal *weight decay* λ which determines the penalization amount of too complex models is found by a grid search over the interval $[0.001, 0.1]$. Similarly, several candidates between 5 and 100 are tested to find the optimal H. One chooses the pair (λ, H) which yields best performance in cross–validation, see Ripley (1996).

3.3. Consistency of convex surrogate loss functions

Membership values generated by the regularization methods described in the two previous sections differ to the ones usually given by statistical classifiers. In regularization membership values are unnormalized scores which do not give any probabilistic confidence about the membership to the corresponding classes while statistical methods generate probabilities that reflect the assessment uncertainty. Such kind of membership probabilities are to be preferred over unnormalized scores for various reasons, see Section 2.2, and therefore unnormalized scores need to be calibrated.

For the regularization methods ANN and SVM calibration means a transformation of unnormalized scores into membership probabilities. Hence, it has to be justified beforehand if the unnormalized classifier scores, at least approximately, supply sufficient information to transform them into probabilities which reflect the assessment uncertainty. In recent years, a connection between the choice of loss function in risk minimization procedures and ability to estimate membership probabilities was investigated, initiated by Lin (1999) and mainly extended by Zhang (2004), Steinwart (2005) and Lugosi & Vayatis (2004).

For classification tasks with a dichotomous class variable $c \in \mathcal{C} = \{-1, +1\}$ regularization methods can be seen as methods which try to find a function $f(\vec{x})$ for which the corresponding classification rule $\hat{c} := \text{sign}[f(\vec{x})]$ minimizes the L–risk $R_L(c, \hat{c})$, see Definition 2.1.6. Since the minimization of the 0–1–risk R_{01} using the obvious 0–1–loss is computationally intractable, see Section 2.4, Machine Learners usually minimize an L–risk using a convex surrogate for L_{01}. Conditioning this L–risk on $\vec{x}$ it can be formulated with incorporating f for the dichotomous case as

$$\begin{aligned} R_L[c, f(\vec{x})] &= E\{L[c \cdot f(\vec{x})]\} \\ &= E\{p_+ L[f(\vec{x})] + [1-p_+] L[-f(\vec{x})]\}, \end{aligned} \quad (3.31)$$

with conditional probability $p_+ := P(+1|\vec{x})$ for class of given observation $\vec{x}$ being positive and complement term $1 - p_+$ as probability for class of given observation $\vec{x}$ being negative. Therefore, this formulation of the risk includes the membership values $f(\vec{x})$ which are generated by the regularization method and the desired membership values, the membership probabilities p_+.

It has been shown by Lin (1999) that the optimal function $f(\vec{x})$ generated by a regularization method which minimizes (3.31) converges in probability to a set–valued function of the conditional probability p_+

$$f_L^*(p_+) := \left\{ g : \mathfrak{R}_L(p_+, g) = \min_{h \in \mathbb{R}} \mathfrak{R}_L(p_+, h) \right\}$$

with *conditional L–risk* $\mathfrak{R}_L(p_+, g) := p_+ L(g) + (1 - p_+) L(-g)$. By symmetry, $\mathfrak{R}_L(p_+, g) = \mathfrak{R}_L(1 - p_+, -g)$ holds. Thus, the optimal function $f_L^*(p_+)$ is not necessarily uniquely determined, since $f_L^*(p_+) = -f_L^*(1 - p_+)$ holds. Additionally, the symmetry of the optimal function implies $f_L^*(0.5) = 0$.

However, $f_L^*(p_+)$ is just a theoretically existing function, but it is not the optimal function $f(\vec{x})$ generated by a classification method which minimizes the L–risk with use of loss function L. $f(\vec{x})$ just approaches $f_L^*(p_+)$ for $N \to \infty$. Since the size of the training set N is usually far from infinity, the conditional probability p_+ can not be simply estimated by $f_L^{*-1}[f(\vec{x})]$ even if f_L^* is bijective. Nevertheless, bijectivity of f_L^* implies that $f(\vec{x})$ supplies information for estimating the conditional probability p_+ on basis of $f(\vec{x})$. Contrariwise, if f_L is not bijective, $f(\vec{x})$ does not contain any probabilistic information about the assessment uncertainty. The differences in observed membership values just occur from approximation and model error. Hence, it is preferable to choose a regularization procedure which contains a convex loss function for which f_L^* is bijective.

For an individual loss function L the particular optimal function $f_L^*(p_+)$ which theoretically minimizes the corresponding risk can be easily found by setting the derivative of (3.31) equal to zero and solve to $f(\vec{x})$. Table 3.1 shows which optimizing functions $f_L^*(p_+)$ the optimal function $f(\vec{x})$ approximates for a corresponding loss function L.

From Table 3.1 it follows that the $L2$–SVM gives more information on estimating conditional probabilities than the $L1$–SVM and is therefore preferred in the experimental analyses in Chapter 6. Furthermore, Table 3.1 shows that besides the statistical methods like LDA and Logistic Regression, see Fisher (1936) and Hosmer & Lemeshow (2000), respectively, the two most common boosting algorithms – *AdaBoost* by Schapire *et al.* (1998) and *LogitBoost* by Friedman *et al.* (2000) – as well as the two standard tree learners – *C4.5* by Quinlan (1993) and *CART* by Breiman *et al.* (1984) – supply sufficient information on estimating conditional probabilities.

Table 3.1.: Risk-minimizing probability functions for different classifiers

Classifier	Loss function	$L(v)$	$f_L^*(p_+)$	Bijective
AdaBoost	Exponential	$\exp(-v)$	$\frac{1}{2}\log\left(\frac{p_+}{1-p_+}\right)$	yes
ANN, CART, LDA	Least Squares	$(1-v)^2$	$2p_+ - 1$	yes
L1–SVM	Hinge	$\max(1-v, 0)$	$\text{sign}(2p_+ - 1)$	no
L2–SVM	Quadratic hinge	$\max(1-v, 0)^2$	$2p_+ - 1$	yes
C4.5, LogitBoost, Logistic Regression	Logistic	$\log(1+e^{-v})$	$\log\left(\frac{p_+}{1-p_+}\right)$	yes

4. Univariate Calibration methods

This chapter gives an overview of the univariate calibration methods which are currently used. In terms of calibration univariate means that the set of possible classes $\mathcal{C}$ consists only of two classes. For convenience these two classes are denoted as a positive and a negative class, respectively, i. e. $k \in \mathcal{C} = \{-1, +1\}$.

The aim of a calibration method is to obtain an estimate for the probability $P(C = k|m)$ for k being the true class by given membership value m. This chapter illustrates four different approaches to provide such calibrated probability estimates:

1. Simple normalization of unnormalized scores, so that they meet mathematical requirements to be a probability, i. e. sum for each observation up to one and lie in the interval $[0, 1]$, see Section 4.1;

2. Estimate a function which maps directly from the membership values $m_+ := m_{\text{method}}(+1|\vec{x})$ (either unnormalized score or membership probability) for the positive class to calibrated probability $\hat{P}_{\text{cal}}(+1|m_+)$. Determine the calibrated probability $\hat{P}_{\text{cal}}(-1|m_+)$ for the negative class by using the complement, see Section 4.2;

3. Estimate the class priors π_k as well as the class–conditional probabilities $P(s_+|C = +1)$ and $P(s_+|C = -1)$ for unnormalized scores to derive calibrated probabilities $\hat{P}_{\text{cal}}(C = k|s_+)$ with Bayes' Rule, see Section 4.3;

4. Regard membership probabilities $p_{i,k} := P_{\text{method}}(C = k|\vec{x}_i)$ (or normalized scores) for the assigned classes as realizations of a Beta distributed random variable and optimize distributional parameters to determine the calibrated probabilities $\hat{P}_{\text{cal}}(C = k|p_{i,k})$, see Section 4.4.

Multivariate Extensions of these univariate calibration methods are presented in Chapter 5.

4.1. Simple Normalization

This section supplies simple normalization procedures for unnormalized scores. Here, these scores are just normalized so that they meet the mathematical requirements to be a probability, i. e. sum for each observation up to one and lie in the interval $[0, 1]$. Furthermore, *boundary values*, e. g. for regularization of unnormalized scores where $s(k|\vec{x}) = 0$, are transformed to boundary membership probabilities with $\hat{P}(k|\vec{x}) = 0.5$.

Such normalized membership values do not cover the assessment uncertainty of an observation belonging to a particular class, though this should be the aim of a calibration method, see Section 2.3. Hence, these methods have no probabilistic background and should only be used for pre–calibration, see Zadrozny & Elkan (2002). Since the calibration method using assignment values, see Section 4.4, needs either membership probabilities or normalized scores as input, one of these simple normalization procedures has to precede a calibrator using assignment values.

Simple Normalization The simplest way of normalization is to divide the observed scores by their range and to add half the range so that boundary values lead to boundary probabilities. Since the boundary in regularization is 0, the range of scores is here equal to the doubled maximum of absolute values of scores. Hence, the following equation

$$\hat{P}_{\text{norm}}(C = k|s_{i,k}) := \frac{s_{i,k} + \rho \cdot \max_{i,j} |s_{i,j}|}{2 \cdot \rho \cdot \max_{i,j} |s_{i,j}|} \ . \tag{4.1}$$

leads to normalized scores which meet the mathematical requirements to be probabilities. The smoothing factor ρ in (4.1) is set to 1.05 if it is necessary to have normalized membership values which are neither exactly equal to 1 nor to 0, otherwise one can set $\rho = 1$.

E–calibration A more sophisticated simple normalization procedure is the following method, called *E–calibration*. Membership probabilities are derived as

$$\hat{P}_{\text{e-cal}}(C = k|s) := \frac{\exp\left[\alpha_k s(C = k|\vec{x})\right]}{\exp\left[\sum_{j=1}^{K} \alpha_j s(C = j|\vec{x})\right]} . \tag{4.2}$$

In this method an estimation of optimal parameters $\alpha_1, \ldots, \alpha_K$ is necessary. The speciality with setting the parameters $\alpha_1, \ldots, \alpha_K = 1$ is called *softmax–calibration*, see Bridle (1989). Nevertheless, E–calibration as well as softmax–calibration have no probabilistic background, but are just (non–)linear transformations. Hence, these methods should not be used for a determination of membership probabilities and will be omitted in the analyses in Chapter 6.

4.2. Calibration via mapping

Calibration with mapping is basically the search for a function which maps a membership value $m_+ := m(C = +1|\vec{x})$ to a calibrated conditional probability $\hat{P}_{\text{cal}}(C = +1|m_+)$ for the positive class. This mapping function can be learned with one out of the various types of regression techniques.

Conditional probabilities for the negative class are usually estimated by using the complement: $\hat{P}_{\text{cal}}(C = -1|m_+) := 1 - \hat{P}_{\text{cal}}(C = +1|m_+)$. Hence, calibration methods which use mapping are regularly only applicable for binary classifier outputs.

All of the following mapping calibrators are applicable for a mapping from either membership probabilities $p_+ := P(C = +1|\vec{x})$ or unnormalized scores $s_+ := s(C = +1|\vec{x})$ to calibrated membership probabilities $\hat{P}_{\text{cal}}(C = +1|m_+)$. The use of unnormalized scores or membership probabilities usually depends on what kind of scores the classifier provides. Without loss of generality, all mapping calibration methods are introduced in the following for a calibration of membership values. The three mapping methods presented in this section are based on Logistic, Piecewise Logistic and Isotonic Regression.

4.2.1. Logistic Regression

The actually most accepted and approved method for the calibration of membership values is to model the log odds of the conditional probabilities

$$\log \frac{\hat{P}_{\text{lr}}(C = +1|m_+)}{\hat{P}_{\text{lr}}(C = -1|m_+)} = g(m_+) \tag{4.3}$$

as a (linear) function g of the membership values for the positive class.
Replacing the probability for the negative class in the log odds (4.3) with using the complement $\hat{P}_{\text{lr}}(C = -1|m_+) = 1 - \hat{P}_{\text{lr}}(C = +1|m_+)$ leads to the term for deriving the calibrated conditional probability

$$\begin{aligned}\hat{P}_{\text{lr}}(C = +1|m_+) &= \frac{\exp\left[g(m_+)\right]}{1 + \exp\left[g(m_+)\right]} \\ &= \frac{1}{1 + \exp\left[-g(m_+)\right]}.\end{aligned} \tag{4.4}$$

A reasonable choice for g, introduced by Platt (1999), is the linear function $g(m) = Am + B$ with scalar parameters A and B. By using this linear function, the calibration function is fitted with a sigmoidal shape, see Figure 4.1.
The search for the mapping function g is an optimization problem. The estimators $\hat{A}$ and $\hat{B}$ are found with the optimization procedure *model–trust algorithm*, see Platt (1999), by minimizing the log–loss error function

$$\mathbf{O}_{lr} := -\sum_{i=1}^{N} \tilde{c}_i \log\left[\hat{P}_{\text{lr}}(+1|m_{i+})\right] + (1 - \tilde{c}_i) \log\left[1 - \hat{P}_{\text{lr}}(+1|m_{i+})\right]$$

with usage of modified *noisy class labels*

$$\tilde{c}_i := \begin{cases} 1 - \epsilon_+ = \dfrac{N_+ + 1}{N_+ + 2} & \text{if } c_i = +1 \\ \epsilon_- = \dfrac{1}{N_- + 2} & \text{''} \quad c_i = -1 \end{cases}$$

where N_+ is the number of positive examples in the training set and N_- is the number of negative ones. These modified noisy class labels are used instead of modified binary class labels $\tilde{c}_i := \mathbf{I}_{[c_i=+1]}(\vec{x}_i) \in \{0, 1\}$ to avoid overfitting.

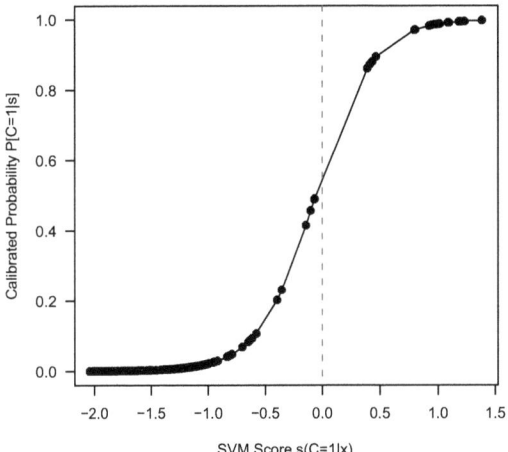

Figure 4.1.: Typical calibration function by using Logistic Regression

4.2.2. Piecewise Logistic Regression

Zhang & Yang (2004) extend Platt's Logistic Regression method by using Piecewise Logistic Regression, while the idea for applying piecewise instead of full Logistic Regression was initiated by Bennett (2002) with his idea of asymmetric distribution of scores, see also Section 4.3.3.

Different from Platt's model (4.3) the log odds are not regarded as a linear function of membership values, but as a piecewise linear function with four knots $(\theta_0 < \theta_1 < \theta_2 < \theta_3)$. Using four knots leads to a separation of membership values into the three areas *obvious decision for negative class* (Area $\mathcal{M}_1$), *hard to classify* (Area $\mathcal{M}_2$) and *obvious decision for positive class* (Area $\mathcal{M}_3$), induced by Bennett (2002), see Section 4.3.3. In each of these three areas the log odds are fitted separately and independently as a linear function, see Figure 4.2.

It is reasonable to chose the minimum and maximum of membership values as estimates for the outer knots $\hat{\theta}_0$ and $\hat{\theta}_3$, respectively. The crucial point is the

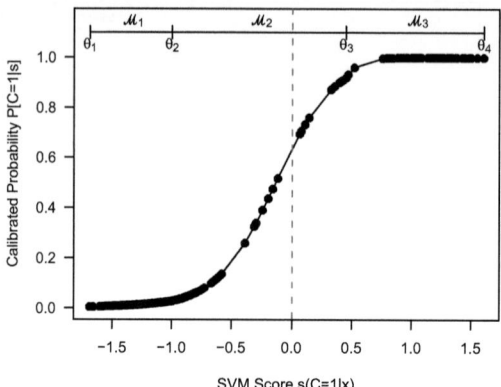

Figure 4.2.: Typical calibration function by using Piecewise Logistic Regression

estimation of inner knots $\hat{\theta}_1$ and $\hat{\theta}_2$ which separate the membership values into the three particular areas. Therefore, the following optimization procedure is repeated by trying any pair of 5%–quantiles of the positive class membership values as candidate pair for the two inner knots. The pair of quantiles which yields best results on the procedure is chosen as the two estimators for the inner knots $\hat{\theta}_1$ and $\hat{\theta}_2$.

As described above, the log odds of membership values (4.3) are modeled as a piecewise linear function

$$g(m_+) = \sum_{j=0}^{3} w_j l_j (m_+) \tag{4.5}$$

with independent weights w_j and independent linear functions

$$l_j(m_+) = \begin{cases} \frac{m_+ - \theta_{j-1}}{\theta_j - \theta_{j-1}} & \text{if } \theta_{j-1} \leq m_+ < \theta_j \quad (j = 1, 2, 3) \\ \frac{m_+ - \theta_{j+1}}{\theta_j - \theta_{j+1}} & \text{''} \quad \theta_j \leq m_+ < \theta_{j+1} \quad (j = 0, 1, 2) \\ 0 & \text{else} \end{cases}$$

Estimating the weight parameters w_j ($j = 0, \ldots, 3$) by Maximum Likelihood is equivalent to minimize the following objective

$$\mathbf{O}_{\text{plr}} := \sum_{j=1}^{3} \sum_{i: m_{i+} \in \mathcal{M}_j} \log \left\{ 1 + \exp\left[-c_i \left(w_j \frac{m_{i+} - \theta_{j-1}}{\theta_j - \theta_{j-1}} + w_{j-1} \frac{m_{i+} - \theta_j}{\theta_{j-1} - \theta_j} \right) \right] \right\}$$

where the partitions $\mathcal{M}_j = \{m_{i+} : \theta_{j-1} \leq m_{i+} < \theta_j, i = 1, \ldots, N\}$ of membership values correspond to the three areas in Figure 4.2.

To avoid overfitting Zhang & Yang (2004) add a regularization term to the objective function

$$\mathbf{O}_{\text{plr},reg} := \mathbf{O}_{\text{plr}} + \lambda \sum_{j=2}^{3} \left(\frac{w_j - w_{j-1}}{\theta_j - \theta_{j-1}} - \frac{w_{j-1} - w_{j-2}}{\theta_{j-1} - \theta_{j-2}} \right)^2,$$

where λ is the regularization coefficient that controls the balance between training loss and model complexity. Zhang & Yang (2004) use $\lambda = 0.001$ in their optimizations.

By using estimated parameters $\hat{w}_j$ in the mapping function $\hat{g}(m_+)$ (4.5) calibrated probabilities are calculated similarly to the Logistic Regression method, described beforehand. Positive class probabiltities

$$\hat{P}_{\text{plr}}(C = +1|m_+) = \frac{1}{1 + \exp[-\hat{g}(m_+)]}$$

are calculated as in (4.4) while negative class probabilities are derived with the complement term $\hat{P}_{\text{plr}}(C = -1|m_+) = 1 - \hat{P}_{\text{plr}}(C = +1|m_+)$.

4.2.3. Isotonic Regression

As extension to *Binning*, see Zadrozny & Elkan (2001a), the following calibration method, introduced by Zadrozny & Elkan (2002), uses *Isotonic Regression* to estimate a function g which describes the mapping from membership values m_{i+} to conditional probabilities $\hat{P}_{\text{ir}}(C = +1|m_{i+})$. Isotonic Regression is a nonparametric form of regression which leads to a stepwise–constant function, see Figure 4.3.

This function which describes the mapping from explanatory to response variable is chosen from the class of all isotonic, i. e. non–decreasing functions. The Isotonic Regression method applies to the calibration problem the basic model

$$\hat{P}_{\text{ir}}(C = +1|m_{i+}) = g(m_{i+}) + \epsilon_i$$

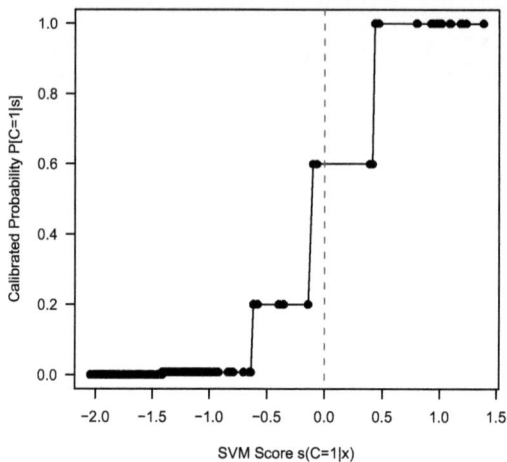

Figure 4.3.: Typical calibration function by using Isotonic Regression

where g is an isotonic function and ϵ_i an individual error term. Given a training set with learned membership values m_{i+} and modified binary class labels

$$\tilde{c}_i = \begin{cases} 1 & \text{if } c_i = +1 \\ 0 & \text{"} \ c_i = -1 \end{cases}$$

a non–decreasing mapping function $\hat{g}$ can be found, so that

$$\hat{g} := \arg\min_{\mathfrak{g}} \sum_{i=1}^{N} [\tilde{c}_i - \mathfrak{g}(m_{i+})]^2 \qquad (4.6)$$

holds. The algorithm *pair–adjacent violators* (PAV), see Algorithm 1, is used to find the stepwise–constant non–decreasing function that best fits the training data according to the mean–square error criterion (4.6).

The PAV–algorithm basically replaces a value $\hat{g}_{i-1}$ and the sequence of values $\hat{g}_i, \hat{g}_{i+1}, \ldots$ which are smaller than $\hat{g}_{i-1}$ by their average. This is shown exemplarily in Figure 4.4.

An application of the PAV–algorithm to a training sample returns a set of intervals and an estimate $\hat{g}_i$ for each interval i, such that $\hat{g}_{i+1} \geq \hat{g}_i$. To obtain a

Algorithm 1 PAV algorithm
1: Input: training set $\{m_{i+}, \tilde{c}_i\}$ sorted with respect to m_{i+}
2: Initialize $\hat{g}_{i,i} = \tilde{c}_i, w_{i,i} = 1$
3: **while** $\exists\ i$ so that $\hat{g}_{k,i-1} > \hat{g}_{i,l}$ **do**
4:    Set $w_{k,l} = w_{k,i-1} + w_{i,l}$
5:    Set $\hat{g}_{k,l} = (w_{k,i-1}\hat{g}_{k,i-1} + w_{i,l}\hat{g}_{i,l})/w_{k,l}$
6:    Replace $\hat{m}_{k,i-1}$ and $\hat{m}_{i,l}$ with $\hat{m}_{k,l}$
7: **end while**
8: Solution: stepwise–constant function generated by $\hat{g}$

class labels $\tilde{c}_i$ (0 1 0 0 1 0 1)

↓ ↓ ↓ ↓

function values $\hat{g}_i$ (0 0.33 0.33 0.33 0.5 0.5 1)

Figure 4.4.: Example for PAV–algorithm

calibrated probability for a test example $\vec{x}$, one finds the interval i in which the generated membership value $m(C = +1|\vec{x})$ falls and assigns the corresponding function value $\hat{g}_i$ as membership probability for the positive class. Thus, the terms for calculating the calibrated membership probabilities become

$$\hat{P}_{\text{ir}}(C = +1|m_+) = \hat{g}(m_+)$$

and $\hat{P}_{\text{ir}}(C = -1|m_{i+}) = 1 - \hat{P}_{\text{ir}}(C = +1|m_{i+})$.

4.3. Calibration via Bayes Rule

While the previously described calibration methods directly map from all kinds of membership values to calibrated probabilities, the following method is only directly applicable for a calibration of unnormalized scores and not for a re–calibration of membership probabilities. Furthermore, in contrast to the direct mapping approaches this method consists of two steps to supply membership probabilities.

At first, the positive class scores s_+ are split into two groups according to their true class, so that probabilities $P(s_+|C=k)$ for the score given a particular class $k \in \{-1,+1\}$ can be derived.

The second step is the determination of membership probabilities by application of Bayes' Theorem to class–conditional probabilities and class priors π_k. While class priors can easily be estimated from the training set, the crucial point in this way of calibration is the choice of the distribution type for the class–conditional probabilities $P(s_+|C=k)$. Two different approaches are presented in this section, the standard assumption of a Gaussian distribution and a further idea using the Asymmetric Laplace distribution.

4.3.1. Idea

Main idea of this method is to estimate the membership probabilities by using Bayes' rule

$$\hat{P}_{\text{bay}}(C=k|s_+) = \frac{\pi_k \cdot P(s_+|C=k)}{\pi_- \cdot P(s_+|C=-1) + \pi_+ \cdot P(s_+|C=+1)}$$

with class priors π_k and class–conditional densities $P(s_+|C=k)$.

Class priors are estimated by calculating class frequencies observed in the training set

$$\hat{\pi}_k := \frac{N_k}{N}, \quad k=-1,+1\,.$$

The estimation of the class–conditional densities is shown in Sections 4.3.2 and 4.3.3 for the Gaussian and the Asymmetric Laplace distribution, respectively.

In estimating the class–conditional densities it is the idea to model the distribution of unnormalized scores instead of membership probabilities. Since some of the standard classification methods, see Hastie et al. (2001), only supply membership probabilities, it is necessary to transform these probabilities to unnormalized scores. Bennett (2002) uses log odds of the probabilistic scores to supply such unnormalized scores

$$s_{\text{method}}(C=+1|\vec{x}) = \log \frac{P_{\text{method}}(C=+1|\vec{x})}{P_{\text{method}}(C=-1|\vec{x})}\,. \qquad (4.7)$$

In calibrating with this method such transformations have to be applied to e. g.
LDA, Naive Bayes or Logistic Regression membership values before calibration.
The unnormalized scores for the positive class, either derived with log odds (4.7)
or directly given, have to separated into two groups. One group consists of the
positive class scores for which $+1$ is the true class

$$\mathcal{S}^+ := \{s_{i+} : c_i = +1, i = 1, \ldots, N\}$$

while the other one contains of the remaining positive class scores for which it is
not

$$\mathcal{S}^- := \{s_{i+} : c_i = -1, i = 1, \ldots, N\} \, .$$

For each group class–conditional densities are derived independently. In the following Sections 4.3.2 and 4.3.3 densities are derived for scores $s \in \mathcal{S}^+$ but derivations work analogously for scores $s \in \mathcal{S}^-$.

4.3.2. Gaussian distribution

In this calibration method the density of scores $s \in \mathcal{S}^+$ has to be estimated. The
standard distributional assumption in statistics is to assume a Gaussian distribution.

Hence, unnormalized scores s are assumed to be realizations of a Gaussian distributed random variable $S \overset{iid}{\sim} G(\mu, \sigma)$ with parameters μ and σ. A Gaussian distributed random variable has got the density function

$$f_G(s|\mu, \sigma) = \frac{1}{\sigma\sqrt{2\pi}} \exp\left[-\frac{1}{2}\left(\frac{s-\mu}{\sigma}\right)^2\right]$$

which has got a symmetric shape, see Figure 4.5.

The parameters of the Gaussian distribution are estimated with Maximum Likelihood by mean and standard deviation of scores:

- $\hat{\mu}^+ := \bar{s}^+ = \dfrac{1}{|\mathcal{S}^+|} \sum_{i:s_i \in \mathcal{S}^+} s_i \, ;$

- $\hat{\sigma}^+ := \sqrt{\dfrac{1}{|\mathcal{S}^+| - 1} \sum_{i:s_i \in \mathcal{S}^+} (\bar{s}^+ - s_i)^2} \, .$

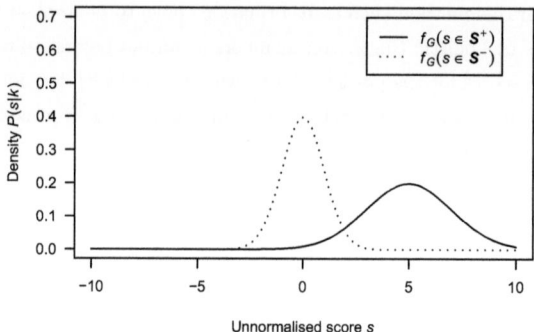

Figure 4.5.: Typical behavior of Gaussian class–conditional densities

Estimation of parameters $\hat{\mu}^-$ and $\hat{\sigma}^-$ for distribution of scores $s \in \mathcal{S}^-$ works analogously.

With estimated class–conditional probabilities

$$\hat{P}_G(s|C=k) := f_G(s \in \mathcal{S}^k | \hat{\mu}^k, \hat{\sigma}^k)$$

and class frequencies as estimators for class–priors

$$\hat{\pi}_k = \frac{N_k}{N}$$

the term for calculating calibrated membership probabilities becomes

$$\hat{P}_{\text{gauss}}(C=k|s) = \frac{\hat{\pi}_k \cdot \hat{P}_G(s|C=k)}{\hat{\pi}_- \cdot \hat{P}_G(s|C=-1) + \hat{\pi}_+ \cdot \hat{P}_G(s|C=+1)}.$$

A typical behavior of a calibration function is similar to the shape of the calibration function for the following calibration with an Asymmetric Laplace distribution, see Figure 4.7.

4.3.3. Asymmetric Laplace distribution

According to Bennett (2002) it is not justifiable to assume for classifier scores a symmetric distribution, e. g. the Gauss distribution as above, but an asymmetric one. He mentions that scores have a different distributional behavior in the area between the modes of the two distributions compared to the respective other side. The area between the modes contains the scores of those observations which are difficult to classify, while the respective other halves stand for the observations for which classification is easier. This conclusion leads to the separation of scores into the three areas *obvious decision for the negative class* (Area A), *hard to classify* (Area B) and *obvious decision for the positive class* (Area C), see Figure 4.6.

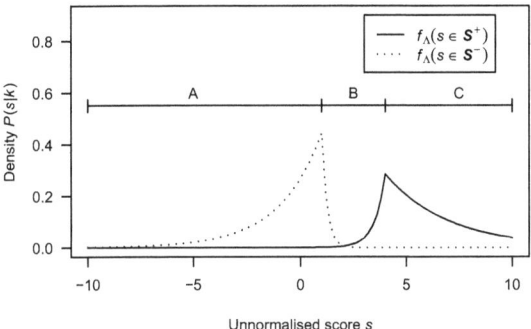

Figure 4.6.: Typical behavior of Asymmetric Laplace class–conditional densities

To consider this different distributional behavior of scores it is required to model the class–conditional densities as asymmetric distributions. Bennett (2002) applies an Asymmetric Gaussian and an Asymmetric Laplace distribution. Since calibration with the Asymmetric Laplace distribution yields better results in his analyses than using the Asymmetric Gaussian, only the calibration with the Asymmetric Laplace distribution $\Lambda(\theta, \beta, \gamma)$ is presented. The Asymmetric Laplace distribution has the density function

$$f_\Lambda(s|\theta, \beta, \gamma) = \begin{cases} \dfrac{\beta\gamma}{\beta+\gamma} \exp\left[-\beta(\theta-s)\right] & \text{if } s \leq \theta \\ \dfrac{\beta\gamma}{\beta+\gamma} \exp\left[-\gamma(s-\theta)\right] & \text{'' } s > \theta \end{cases}$$

with scale parameters $\beta, \gamma > 0$ and mode θ. Scale parameter β is the inverse scale of the exponential belonging to the left side of the mode, while γ is the inverse scale of the exponential belonging to the right side.

In estimating the parameters for distribution of the scores $s \in \mathcal{S}^+$ with given training set $\vec{x}_1, \ldots, \vec{x}_N$ the unnormalized scores s are assumed to be realizations of an Asymmetric Laplace distributed random variable $S \stackrel{iid}{\sim} \Lambda(\theta, \beta, \gamma)$ with likelihood function

$$l(\theta, \beta, \gamma) = \prod_{s_i \in \mathcal{S}^+} f_\Lambda(s_i | \theta, \beta, \gamma). \qquad (4.8)$$

For estimation of scale parameters β and γ the following procedure is repeated for every score realization s_i as candidate for mode θ. The corresponding estimators $\hat{\beta}_\theta$ and $\hat{\gamma}_\theta$ are evaluated for each candidate by Maximum Likelihood. Finally, the candidate θ and corresponding estimators $\hat{\beta}_\theta$ and $\hat{\gamma}_\theta$ which attain highest likelihood are chosen as parameter estimators.

In the beginning of the estimation procedure the $N^+ := |\mathcal{S}^+|$ learned scores are separated into two groups with scores lower/higher than candidate θ:

$$\mathcal{S}_l = \left\{s_i : s_i \leq \theta; i = 1, \ldots, N^+\right\} \quad ; \quad \mathcal{S}_r = \left\{s_i : s_i > \theta; i = 1, \ldots, N^+\right\}$$

with potencies $N_l := |\mathcal{S}_l|$ and $N_r := |\mathcal{S}_r|$.

Afterwards the sum of absolute differences between θ and the s_i belonging to the left/right halve of the distribution is calculated:

$$D_l = N_l \theta - \sum_{i : s_i \in \mathcal{S}_l} s_i \quad ; \quad D_r = \sum_{i : s_i \in \mathcal{S}_r} s_i - N_r \theta.$$

For a fixed θ the Maximum Likelihood estimates for the scale parameters are

$$\hat{\beta}^+ = \frac{N^+}{D_l + \sqrt{D_r D_l}} \quad ; \quad \hat{\gamma}^+ = \frac{N^+}{D_r + \sqrt{D_r D_l}}.$$

Finally, the candidate θ which maximizes the likelihood (4.8) is chosen as mode estimator $\hat{\theta}^+$ with corresponding estimators $\hat{\beta}^+$ and $\hat{\gamma}^+$.

The estimation of scale parameters and mode for the class–conditional distribution of scores $s \in \mathcal{S}^-$ works analogously with using all observed scores as candidates for the same procedure.

As described before, the calibrated probabilities are calculated by using Bayes' rule

$$\hat{P}_{\text{alap}}(C = k | s) = \frac{\hat{\pi}_k \cdot \hat{P}_\Lambda(s | C = k)}{\hat{\pi}_- \cdot \hat{P}_\Lambda(s | C = -1) + \hat{\pi}_+ \cdot \hat{P}_\Lambda(s | C = +1)}.$$

with class–conditional densities

$$\hat{P}_\Lambda(s|C=k) := f_\Lambda\left(s \in \mathcal{S}^k | \hat{\theta}^k, \hat{\beta}^k, \hat{\gamma}^k\right)$$

based on the estimated distributional parameters and smoothed class frequencies as estimators for class–priors

$$\hat{\pi}_k = \frac{N_k + 1}{N + 2}$$

where N_k is the number of examples in class k.
Using the Bayes method with the Asymmetric Laplace distribution leads to a calibration function which can be split into three parts see Figure 4.7.

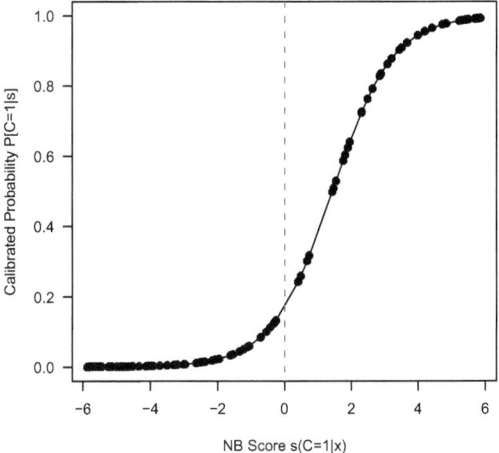

Figure 4.7.: Typical calibration function by using Asymmetric Laplace densities

Each part in the calibration function represents one area of the class–conditional density plot, see Figure 4.6. The two areas with the extreme probabilities correspond to the examples which are easy to classify i. e. Area A and C. In Figure 4.7 that is the line with points going from -6 to -3 for the negative class and the line with points going from approximately 5 to 6 for the positive class. Finally, the curve connecting those two areas represents the observations which are difficult to classify, i. e. Area B.

4.4. Calibration by using assignment values

Calibration methods based on Bayes' rule, see Section 4.3, consist of partitioning and separate calibration. The partitioning is the basis for the following independent determination of calibrated membership probabilities. While the previous calibration method by Bennett (2002) partitions the unnormalized scores for a chosen class according to their true class, Garczarek (2002) partitions the membership values $m(k|\vec{x})$ according to their assignment

$$\hat{c}(\vec{x}_i) := \arg\max_k m(C = k|\vec{x}_i)$$

instead. For a classification problem with number of class labels $k = 1, 2$ the training sample tuple $\mathcal{T} = \{(c_i, \vec{x}_i) : i = 1, \ldots, N\}$ is split into partitions

$$\mathcal{T}_k := \{(c_i, \vec{x}_i) \in \mathcal{T} : \hat{c}(\vec{x}_i) = k\}$$

with potencies $N_{\mathcal{T}_k} = |\mathcal{T}_k|$.

The idea of this method is to model the membership values for the assigned classes in each partition separately as Beta distributed random variables. Therefore, unnormalized scores generated by a regularization method, e. g. the SVM, have to be normalized or "pre–calibrated" with a simple normalization method, see Section 4.1, so that scores sum up to one and lie in the interval $[0, 1]$.

The calibration procedure transforms these normalized membership values $P(k|\vec{x})$ for each partition to new Beta random variables $\hat{P}_{av}(k|\vec{x})$ with optimal parameters and regards them as membership probabilities. Since such calibrated probabilities should cover the assessment uncertainty of the classifier, the correctness rate in the corresponding partition is regarded in this transformation.

Transforming distributional parameters of a random variable can be easily done by using one of the main theorems in statistics, see e. g. Hartung et al. (2005).

Theorem 4.4.1 (Fundamental property of the distribution function)
If Y is a random variable with continuous distribution function F_Y, then the random variable

$$U := F_Y(Y) \sim U[0, 1]$$

is uniformly distributed on the interval $[0, 1]$.

For a uniformly distributed random variable $U \sim [0, 1]$ and any continuous distribution function $F_\mathcal{D}$ of a distribution $\mathcal{D}$ it is true that

$$F_\mathcal{D}^{-1}(U) \sim \mathcal{D}.$$

The Gamma and the Beta distribution are essential for the actual calibration method presented in Section 4.4.1, see Sections A.1.1 and A.1.2 in the appendix for details on these distributions.

4.4.1. Calibration by inverting the Beta distribution

As described above, this calibration method requires the determination of membership probabilities for the assigned classes

$$a_{i,k} := \begin{cases} \max_{k \in \mathcal{C}} P(C = k | \vec{x}_i) & \text{if } \hat{c}(\vec{x}_i) = k \\ \text{not defined} & \text{else} \end{cases} \quad (4.9)$$

which will be called *assignment values* in the following. For each partition these assignment values are assumed to be realizations of a Beta distributed random variable $A_k \sim \mathcal{B}(p_{A_k}, N_{A_k})$ with unknown parameters $p_{A_k} \in [0, 1]$ and $N_{A_k} \in \mathbb{N}$. These two distributional parameters can be estimated by the method of moments:

$$\hat{p}_{A_k} := \bar{a}_k$$
$$\hat{N}_{A_k} := \frac{\bar{a}_k(1 - \bar{a}_k)}{S_k} - 1$$

with moments

$$\bar{a}_k := \frac{1}{N_{\mathcal{T}_k}} \sum_{i: \vec{x}_i \in \mathcal{T}_k} a_{i,k}$$
$$S_k := \frac{1}{N_{\mathcal{T}_k} - 1} \sum_{i: \vec{x}_i \in \mathcal{T}_k} (a_{i,k} - \bar{a}_k)^2.$$

Additionally, Garczarek (2002) introduces the parameter correctness probability φ^k which is also seen for each partition separately as a Beta distributed random variable $\varphi^k \sim \mathcal{B}(p_{\mathcal{T}_k}, N_{\mathcal{T}_k})$. The expected value parameter is quantified by the local correctness rate

$$p_{\mathcal{T}_k} := \frac{1}{N_{\mathcal{T}_k}} \sum_{i : c_i \in \mathcal{T}_k} \mathbf{I}_{[\hat{c}(\vec{x}_i) = c_i]}(\vec{x}_i) \qquad (4.10)$$

and the dispersion parameter is quantified by the number $N_{\mathcal{T}_k}$ of examples in the corresponding partition $\mathcal{T}_k$.

Since calibrated probabilities should reflect the uncertainty about their assignments, assignment values $a_{i,k}$ are transformed from Beta variables with expected value p_{A_k} to Beta variables with local correctness rate $p_{\mathcal{T}_k}$ as expected value. With using Theorem 4.4.1 such transformation can be easily done so that calibrated assignment values become

$$\hat{P}_{\text{av}}(C = k | a_{i,k}) := F^{-1}_{\mathcal{B}, p_{\mathcal{T}_k}, N_{k,opt}} \left[F_{\mathcal{B}, \hat{p}_{A_k}, \hat{N}_{A_k}}(a_{i,k}) \right] . \qquad (4.11)$$

With old parameters $\hat{p}_{A_k}$ and $\hat{N}_{A_k}$ as well as new parameter $p_{\mathcal{T}_k}$ already quantified, it is only required to find an optimal new dispersion parameter $N_{k,opt}$ for calibrating probabilities. The optimal parameter $N_{k,opt}$ found by the *Assignment Value Algorithm*, see Algorithm 2, is the integer N out of the interval $\{N_{\mathcal{T}_k}, N_{A_k}\}$ which maximizes the following objective

$$\mathbf{O}_{\text{av}} := N_{\mathcal{T}_k} \cdot p_{\mathcal{T}_k} + \mathbf{Ac} . \qquad (4.12)$$

$\mathbf{O}_{\text{av}}$ counts the number of correctly assigned examples regularized with the performance measure accuracy

$$\mathbf{Ac} := 1 - \frac{K}{K-1} \frac{1}{N} \sum_{i=1}^{N} \sqrt{\sum_{k=1}^{K} \left[\mathbf{I}_{[c_i = k]}(\vec{x}_i) - P_N(C = k | s) \right]^2} ,$$

to avoid overfitting. This measure is equivalent to the **RMSE** (2.4), see Section 2.4.4.

The Assignment Value Algorithm, see Algorithm 2, is used for each partition separately to supply calibrated probabilities.

Since modeling and calibration works independently for both partitions the calibration function usually has got a jump from negative to positive scores, see Figure 4.8.

One can see in the example of Figure 4.8 that calibration with assignment values leads to two independent non–decreasing functions.

Algorithm 2 Assignment Value Algorithm

1: **for** all integer $N \in \{N_{\mathcal{T}_k}, N_{A_k}\}, N \in \mathbb{N}$ **do**
2:     Estimate calibrated assignment values

$$\hat{P}_{\text{av},N}(C = k | a_{i,k}) = F^{-1}_{\mathcal{B}, p_{\mathcal{T}_k}, N}\left[F_{\mathcal{B}, p_{A_k}, N_{A_k}}(a_{i,k})\right]$$

3:     Determine calibrated probabilities for the class which is not assigned to by using the complement

$$\hat{P}_{\text{av},N}(C \neq k | a_{i,k}) = 1 - \hat{P}_{\text{av},N}(C = k | a_{i,k})$$

4: **end for**
5: As $N_{k,opt}$ choose the N and corresponding probabilities

$$\hat{P}_{\text{av}}(C = k | a_{i,k}) := \hat{P}_{\text{av},N_{k,opt}}(C = k | a_{i,k})$$
$$\hat{P}_{\text{av}}(C \neq k | a_{i,k}) := \hat{P}_{\text{av},N_{k,opt}}(C \neq k | a_{i,k})$$

which minimize the objective $\mathbf{O}_{\text{av}}$ (4.12)

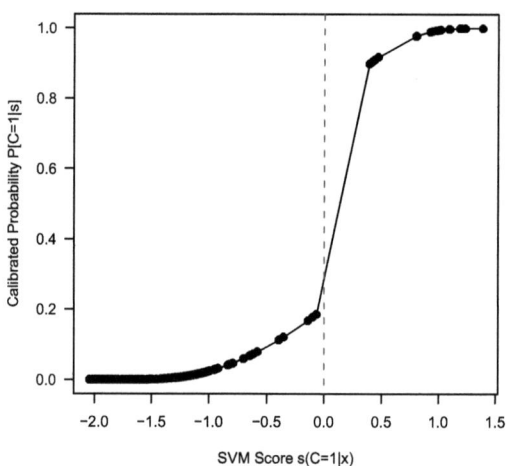

Figure 4.8.: Typical calibration function by using assignment values

5. Multivariate Extensions

This chapter deals with the learning of classification rules for *polychotomous* or *multi–class problems*, i. e. for tasks that cover more than two classes. Therefore, Section 5.1 gives a short overview about how regularization–based and other classification methods solve such multi–class problems. Although many classification problems are multi–class ones, the Support Vector Machine, see Chapter 3, is only applicable for the *binary* case, since this method is based on maximizing the margin between just two classes. For the multi–class case the common approach in regularization, especially for the SVM but also applied to other methods, is a two–step method that transfers the polychotomous classification task into several binary classification problems and combines the binary output after the learning. A framework which unifies the common methods for reducing multi–class into binary is given by Allwein *et al.* (2000). This framework will be presented in Section 5.2. For calibration with subsequent combination of the binary task outcomes into multi–class membership probabilities Zadrozny (2001) gives a general approach based on the pairwise coupling method by Hastie & Tibshirani (1998), see Section 5.3. Additionally, a new one–step multivariate calibration method with using the Dirichlet distribution will be presented in Section 5.4. A detailed experimental comparison of these multivariate extensions methods follows in Chapter 6.

5.1. Procedures in Multi–class classification

A slight overview about standard classification methods, based on Statistics as well as on Machine Learning, and some of its major properties is given with Table 5.1. This overview includes the kind of membership values the various classifiers generate and how these methods deal with multi–class problems.

Table 5.1.: Overview on classification methods

Classifier	Membership values	Multi-class tasks	Assumptions/Idea	Problems/Issues
ANN	Unnormalized Scores	One–against–rest / Direct	Perceptron structure	Model selection, Choice of size and number of layers
Boosting	Unnormalized Scores	One–against–rest	Combination of weak learners	Model selection, Learner and loss function choice
LDA / QDA	Membership Probabilities	Direct	Multivariate Normality with equal/differing covariance matrix	Multivariate Normality not necessarily appropriate
Logistic Regression	Membership Probabilities	Direct	Model the log odds of probabilities as linear function	Linearity of log odds not guaranteed
Naive Bayes	Membership Probabilities	Direct	Independence of feature variables	Membership Probabilities tend to be too extreme due to lack of independence
SVM	Unnormalized Scores	One–against–rest / All–pairs	Linear Separation in high–dimensional feature space	Model selection, Kernel choice
Trees	Proportions regarded as Membership Probabilities	Direct	Linear Partition of feature space	Proportions tend to be too extreme, since Tree methods try to create pure trees

The regularization-based classifiers Boosting, see Schapire *et al.* (1998), ANN and SVM apply or even require binary reduction algorithms for multi-class tasks and generate unnormalized scores which have got no probabilistic background. Instead, the classifiers Logistic Regression, Naive Bayes and Discriminant Analysis, see e. g. Hastie *et al.* (2001) for details on these methods, which all base on statistical theory create membership probabilities and are directly applicable for K-class problems. Tree learners which were initially introduced in statistics, see CART algorithm by Breiman *et al.* (1984), directly generate proportions that are considered as membership probabilities.

Additionally, Table 5.1 presents the major idea of the classification methods and some key issues which occur in the learning of the rules or in the estimation of the membership probabilities. Probabilities generated by the Naive Bayes and Tree classifiers tend to be too extreme which has been explored by Domingos & Pazzani (1996) and Zadrozny & Elkan (2001b), respectively. Membership probabilities generated by statistical classifiers like Logistic Regression or LDA and QDA base on the assumptions these methods make. Calculated probabilities might be inappropriate, if these assumptions fail.

In contrast to the assumption-based statistical classifiers, regularization methods try to find the optimal solution under pre-selected model conditions. All these methods have different opportunities of model selection that influence the outcome, size and number of layers for ANN, learner in Boosting and Kernel function for SVM. Additionally, Section 3.3 discusses the implication of the chosen loss function. These vast opportunities of model selection in regularization lead to the fact that there is no clear and direct way to the desired outcome of membership values. For every particular classification problem the model has to be adjusted. The differing ways of supplying K-class membership probabilities for polychotomous classification tasks are presented in Figure 5.1.

The fastest and apparently easiest approach is the usage of a direct multi-dimensional classifier like LDA or Naive Bayes. Anyway, these methods perform not necessarily best, due to the miscalculations when assumptions are not met, see Table 5.1. As described in Chapter 3, this direct approach is not possible for the SVM classifiers and a binary reduction with one of the methods of Section 5.2 is obedient in this case.

Hence, in K-class situations SVM methods require an algorithm which combines

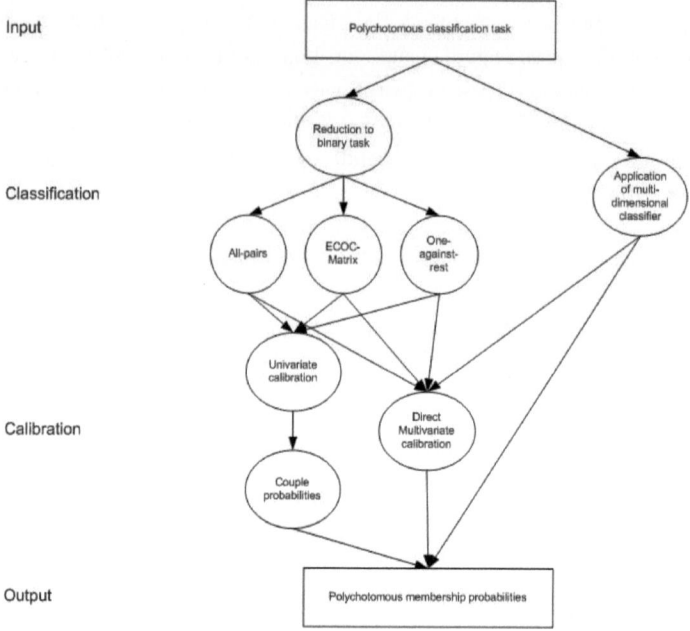

Figure 5.1.: Multivariate Calibration Scheme

the scores generated in the binary classification problems and calibrates them into membership probabilities for the K classes. Currently, the pairwise coupling algorithm by Hastie & Tibshirani (1998) and its extension by Zadrozny (2001) are the common approaches for this problem, see Section 5.3.

Compared to this pairwise coupling with univariate calibration the new Dirichlet calibration method which will be introduced in Section 5.4 has got the advantage, that it conducts just one optimization procedure instead of two. Additionally, this flexible method is applicable to outcomes of binary reduction approaches as well as directly applicable to multivariate matrix outcomes.

5.2. Reduction to binary problems

To generate membership values of any kind, the SVM needs to transfer multi–class tasks into several binary classification problems. The common methods for reducing multi–class into B binary problems are the *one–against rest* and the *all–pairs* approach. Allwein *et al.* (2000) generalize these ideas with using so–called *error–correcting output coding (ECOC)* matrices, introduced by Dietterich & Bakiri (1995). Allwein *et al.* (2000) further introduce reduction approaches with the *sparse* and the *complete ECOC–matrix* which are further dealt with in Section 5.2.2 and 5.2.3, respectively.

All the reduction algorithms have in common that in each learning of a binary rule $b \in \{1, \ldots, B\}$ observations of one ore more classes are considered as positive class observations while observations of another class (group) are treated as negative class observations.

The way classes are considered in a particular binary task b is incorporated into a code matrix Ψ with K rows and B columns. Each column vector $\vec{\psi}_b$ determines the class treatment for the bth classification task with its elements $\psi_{k,b} \in \{-1, 0, +1\}$. A value of $\psi_{k,b} = 0$ implies that observations of the respective class k are ignored in the current task b while values of -1 and $+1$ determine whether a class is regarded as negative and positive class, respectively.

5.2.1. Common reduction methods

This section presents the two widely used and recommended methods, see e. g. Vapnik (2000) or Schölkopf *et al.* (1995), for a reduction to binary classification tasks, the one–against rest and the all–pairs approach. The approaches are generalized in notation of the ECOC–matrix idea by Allwein *et al.* (2000).

One–against rest approach In the one–against rest approach the number of binary decisions B is equal to the number of classes K. Each class is considered as positive once while all the remaining classes are labelled as negative. Hence, the resulting code matrix Ψ is of size $K \times K$, displaying $+1$ on the diagonal while all other elements are -1. Table 5.2 shows an example of a one–against rest matrix Ψ for four classes.

Table 5.2.: Example of a one–against rest code matrix Ψ for four classes

Class	Task 1	2	3	4
1	+1	−1	−1	−1
2	−1	+1	−1	−1
3	−1	−1	+1	−1
4	−1	−1	−1	+1

All–pairs approach In the application of the all–pairs approach one learns a rule for every single pair of classes. In each binary task b one class is considered as positive and the other one as negative. Observations which do not belong to either of these classes are omitted in the respective learning procedure. Thus, Ψ is a $K \times \binom{K}{2}$–matrix with each column b consisting of elements $\psi_{k_1,b} = +1$ and $\psi_{k_2,b} = -1$ corresponding to a distinct class pair (k_1, k_2) while all the remaining elements are 0, as shown exemplary for four classes in Table 5.3.

Table 5.3.: Example of an all–pairs code matrix Ψ for four classes

Class	Task 1	2	3	4	5	6
1	+1	+1	+1	0	0	0
2	−1	0	0	+1	+1	0
3	0	−1	0	−1	0	+1
4	0	0	−1	0	−1	−1

In comparing the all–pairs to the one–against rest approach the all–pairs matrix has got on the one hand more columns and hence more binary rules to learn. On the other hand, the one–against rest matrix does not include any zeros while for four classes half of the elements of the all–pairs matrix are zeros and with increasing number the part of zeros rises, too. This leads to the fact that in using all–pairs one has to learn rules for more problems, but these rules are a lot faster to learn, since they are all learned on the basis of at most just half of the observations.

5.2.2. Sparse ECOC–matrix

Additionally, Allwein et al. (2000) present another reduction approach, the usage of the so called *sparse ECOC–matrix* with $B = 15\log_2(K)$ columns/ classification tasks where all elements are chosen randomly and independently. An element $\psi_{k,b}$ is drawn as 0 with probability 0.5 and drawn as -1 or $+1$ with probability 0.25 each. 10000 matrices of this kind are created and out of all *valid* matrices, the matrices which neither contain identical columns nor columns without a -1 or a $+1$, the one is chosen for which the *Hamming distance*

$$d_H\left(\vec{\psi}_{k_1}, \vec{\psi}_{k_2}\right) = \sum_{b=1}^{B} \left(\frac{1 - \text{sign}(\psi_{k_1,b}\psi_{k_2,b})}{2}\right)$$

is minimal. Dietterich & Bakiri (1995) mention that by using the matrix with minimal Hamming distance the performance in learning multi–class problems is improved significantly and robustly, since the shape of the ECOC–matrix with minimal d_H enhances the correction of errors in the binary decisions. Anyway, some problems which occur in the construction of such sparse ECOC–matrices will be shown in the following.

Computational properties Idea of this approach is to generate a matrix which on the one hand is sparse in the sense that B is smaller than for all–pairs and has therefore less binary tasks to learn rules for, but that on the other hand still sufficiently separates between the classes. However, the sparse ECOC–matrix creation algorithm which is introduced by Allwein et al. (2000) has got a major drawback. Before choosing the sparsest matrix 10000 different matrices are generated randomly. On the one hand, the computation of such a number of matrices takes quite long, if the number of classes K is high. On the other hand, if K is small, even these 10000 repeats do not necessarily suffice to supply at least one matrix that is valid. The mathematical proof for this statement will be shown in the following. Therefore, the probability for the creation of a valid matrix is derived on the basis of combinatorics. This probability is the groundwork for the subsequent derivation of the probability that at least one out of the 10000 generated matrices is valid.

At first, consider the probability $\mathcal{P}_{-,+} := P\left(-1 \in \vec{\psi}_b \wedge +1 \in \vec{\psi}_b\right)$ that the bth column of the code matrix Ψ contains at least one $+1$ and one -1. This probability can be regarded as the complement of the probability that the column vector $\vec{\psi}_b$ does not include a single $+1$ or not a single -1:

$$\begin{aligned}
\mathcal{P}_{-,+} &= 1 - P\left(-1 \notin \vec{\psi}_b \vee +1 \notin \vec{\psi}_b\right) \\
&= 1 - \left[P\left(+1 \notin \vec{\psi}_b\right) + P\left(-1 \notin \vec{\psi}_b \wedge +1 \in \vec{\psi}_b\right)\right] \\
&= 1 - P\left(+1 \notin \vec{\psi}_b\right) - P\left(-1 \notin \vec{\psi}_b \wedge +1 \in \vec{\psi}_b\right) .
\end{aligned} \quad (5.1)$$

To solve the equation for $\mathcal{P}_{-,+}$ (5.1), the required probabilities $P\left(+1 \notin \vec{\psi}_b\right)$ and $P\left(-1 \notin \vec{\psi}_b \wedge +1 \in \vec{\psi}_b\right)$ will be derived in the following.

Since the column vector $\vec{\psi}_b$ has length K and the probability for occurrence of $+1$ is 0.25,

$$P\left(+1 \notin \vec{\psi}_b\right) = (1 - 0.25)^K = (0.75)^K \quad (5.2)$$

holds for the probability that not even one $+1$ is included in the column vector. The remaining probability $P\left(-1 \notin \vec{\psi}_b \wedge +1 \in \vec{\psi}_b\right)$ that the column vector contains at least one $+1$, but no -1 can be determined by

$$P\left(-1 \notin \vec{\psi}_b \wedge +1 \in \vec{\psi}_b\right) = P\left(-1 \notin \vec{\psi}_b\right) - P\left(-1 \notin \vec{\psi}_b \wedge +1 \notin \vec{\psi}_b\right) \quad (5.3)$$

with $P\left(-1 \notin \vec{\psi}_b\right) = (0.75)^K$ as in Equation (5.2). Furthermore, the latter probability that neither a $+1$ nor a -1 is included in the column vector is

$$\begin{aligned}
& P\left(-1 \notin \vec{\psi}_b \wedge +1 \notin \vec{\psi}_b\right) \\
&= P\left[\vec{\psi}_b = (0, \ldots, 0)'\right] = (0.5)^K,
\end{aligned}$$

since this probability is equal to the probability that the column consists of zeros only and these are drawn with probability 0.5.

With incorporating the derived probabilities from Equations (5.2) – (5.3) into Equation (5.1) one gets the probability that the bth column of the code matrix Ψ contains at least one $+1$ and one -1 with

$$\begin{aligned}
\mathcal{P}_{-,+} &= 1 - P\left(+1 \notin \vec{\psi}_b\right) - P\left(-1 \notin \vec{\psi}_b\right) + P\left(-1 \notin \vec{\psi}_b \wedge +1 \notin \vec{\psi}_b\right) \\
&= 1 - (0.75)^K - (0.75)^K + (0.5)^K \\
&= 1 - 2 \cdot (0.75)^K + (0.5)^K .
\end{aligned} \quad (5.4)$$

Figure 5.2 presents this probability as a function of the number of classes K. Naturally, with increasing K the probability that at least one $+1$ and one -1 is

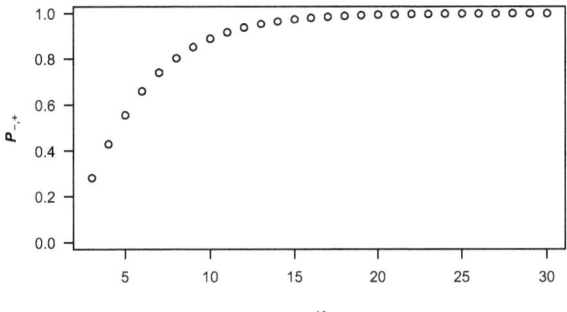

Figure 5.2.: Probability that ECOC–matrix column contains at least one -1 & one +1

in the column vector increases as well.

However, in creation of a sparse code matrix the drawing of a column vector $\vec{\psi}_b$ is repeated B times. This implicates that $(\mathcal{P}_{-,+})^B$ is the probability for the drawing of a matrix Ψ in which every column contains at least one -1 and one $+1$.

Anyway, an additional requirement for a valid code matrix is that no column vector $\vec{\psi}_b$ exists more than once in the matrix, since a doubled vector would lead to a repetition of a binary classification problem. A valid column vector is drawn at least with probability $(1/4)^K$ if it only consists of -1 and $+1$. Since it would lead to an invalid matrix if a vector is drawn again, the maximum probability of a valid code matrix becomes

$$\begin{aligned}\mathcal{P}_{-,+,B} &:= P\left(-1 \in \vec{\psi}_b \wedge +1 \in \vec{\psi}_b \wedge \vec{\psi}_b \neq \vec{\psi}_a | a,b = 1,\ldots,B; a \neq b\right) \\ &\leq \mathcal{P}_{-,+} \cdot \left(\mathcal{P}_{-,+} - \frac{1}{4^K}\right) \cdot \ldots \cdot \left(\mathcal{P}_{-,+} - (B-1) \cdot \frac{1}{4^K}\right).\end{aligned} \quad (5.5)$$

To yield the probability that none of the 10000 created matrices is valid, one can consider the complement probability $1 - \mathcal{P}_{-,+,B}$ that at least one generated ECOC–matrix is valid. Hence, the probability that at least one of the 10000 matrices is valid is given by

$$\mathcal{P}_{-,+,B,10000} = 1 - (1 - \mathcal{P}_{-,+,B})^{10000}$$

This probability is the upper bound for the probability of creating a valid code matrix and is shown in Figure 5.3 as a function of the number of classes K.

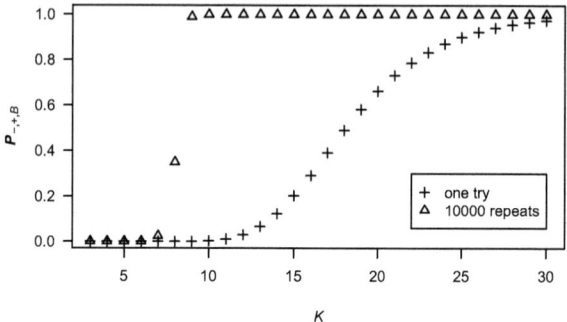

Figure 5.3.: Probability for a valid ECOC–matrix with one try or 10000 repeats

Figure 5.3 demonstrates that the use of a sparse ECOC–matrix is only reasonable for classification tasks with at least nine classes, since for K equal to eight the probability is smaller than 0.4 and for K less than eight it is approximately 0. As already stated beforehand, for large number of classes the creation of 10000 ECOC–matrices takes reasonable long time.

Since this method is not generally applicable in classification, it is omitted in the analyses of Chapter 6, especially because most of the classification tasks cover less than nine classes.

5.2.3. Complete ECOC–matrix

The last presented ECOC procedure is the use of a *Complete* ECOC–matrix, see Allwein *et al.* (2000), which includes all valid combinations of $+1$, 0 and -1 as columns, see the Example in Table 5.4.

Here, one has to keep in mind that in terms of learning a binary classification rule every column has a "twin", e. g. the classification task for column $(+1, -1, 0)'$ is

Table 5.4.: Example of a complete ECOC matrix $\boldsymbol{\Psi}$ for three classes

Class	\multicolumn{6}{c}{Task}					
	1	2	3	4	5	6
1	+1	+1	0	+1	−1	+1
2	−1	0	+1	+1	+1	−1
3	0	−1	−1	−1	+1	+1

identical to the one for column $(-1, +1, 0)'$, since both binary tasks would lead to a rule and membership values which just differ by their algebraic sign. There exists just one unique vector without a twin, the one which consists only of zeros. While the elaboration of all valid out of all 3^K possible combinations is rather simple for three classes it gets more and more complicated with increasing number of classes, which will be shown in the following derivations.

Nevertheless, the number v_K of valid combinations and hence the number B of binary decisions for a complete ECOC–matrix can be derived by using iteration. From Table 5.4 it follows that $v_3 = 6$ is the starting point. If you want to derive the number v_{K+1} of valid columns for $K+1$ classes you have to regard the v_K valid and the $3^K - v_K$ invalid columns for the K classes separately. Easily seen, any column of the v_K valid columns can be extended in the additional row $K+1$ with either a $+1$, a 0 or a -1 which leads to $3 \cdot v_K$ non–identical valid columns, see the example in Figure 5.4.

$$\begin{pmatrix} 0 \\ -1 \\ +1 \end{pmatrix} \rightarrow \begin{pmatrix} 0 \\ -1 \\ +1 \\ +1 \end{pmatrix}, \begin{pmatrix} 0 \\ -1 \\ +1 \\ -1 \end{pmatrix}, \begin{pmatrix} 0 \\ -1 \\ +1 \\ 0 \end{pmatrix}$$

Figure 5.4.: Example for extending a valid ECOC–column for three classes

Furthermore, the $3^K - v_K$ invalid columns for the K classes can be extended to more valid columns by adding either a -1 or a $+1$ in row $K+1$, except just the one row which only contains zeros. This would lead to $3^K - v_K - 1$ additional valid columns, exemplary shown in Figure 5.5.

Anyway, for the reasons mentioned in the beginning of this section, every one of the $3^K - 1$ combinations occurs twice just with inverted algebraic signs. Hence,

$$\begin{pmatrix} 0 \\ -1 \\ 0 \end{pmatrix} \rightarrow \begin{pmatrix} 0 \\ -1 \\ 0 \\ +1 \end{pmatrix} ; \begin{pmatrix} 0 \\ 0 \\ +1 \end{pmatrix} \rightarrow \begin{pmatrix} 0 \\ 0 \\ +1 \\ -1 \end{pmatrix}$$

Figure 5.5.: Example for extending an invalid to a valid ECOC–column for three classes

the number of class K invalid columns that become valid with adding row $K+1$ is

$$3^K - v_K - 1 - \frac{1}{2}\left(3^K - 1\right) = \frac{1}{2}\left(3^K - 1\right) - v_K .$$

Thus, the number of valid columns can be derived by the iterative equation

$$v_{K+1} = 3 \cdot v_K + \frac{1}{2} \cdot \left(3^K - 1\right) - v_K .$$

5.2.4. Comparison of ECOC–matrix based reduction approaches

Table 5.5 compares the required number of binary decisions for the four presented ECOC methods.

The comparison of B values in Table 5.5 clarifies that the application of the complete matrix procedure is not reasonable for a number of classes K greater than 5. Additionally, Allwein et al. (2000) mention that this method is computationally intractable with a number of classes K greater than 13. Thus, this procedure is also omitted in the following experiments, since it is unreasonably more time–intensive than the other reduction approaches.

Table 5.5.: Number of Binary decisions for ECOC reduction approaches

K	One–against rest	All–pairs	Sparse	Complete
3	3	3	24	6
4	4	6	30	25
5	5	10	35	90
6	6	15	39	301
7	7	21	42	966
8	8	28	45	3025
9	9	36	48	9330
10	10	45	50	28501
11	11	55	52	86526
12	12	66	54	261625
13	13	78	56	788970
14	14	91	57	2375101
15	15	105	59	7141686
16	16	120	60	21457825
17	17	136	61	64439010
18	18	153	63	193448101
19	19	171	64	580606446
20	20	190	65	1742343625

5.3. Coupling probability estimates

By coupling probability estimates Hastie & Tibshirani (1998) supply a method for estimating calibrated membership probabilities in cooperation with the all–pairs approach. Zadrozny (2001) extends this method for the application of outcomes generated by a reduction based on any type of ECOC–matrix.

As described before, the reduction approaches apply with each column $\vec{\psi}_b$ of matrix Ψ as class code a binary classification procedure to the data set. Hence, the output of the reduction approach consists of B score vectors $\vec{s}_{+,b}(\vec{x}_i)$ for the associated (group of) positive class(es). To each set of scores separately one of the univariate calibration methods, described in Chapter 4, can be applied. The outcome is a calibrated membership probability $\vec{p}_{+,b}(\vec{x}_i)$ which reflects the probabilistic confidence in assessing observation $\vec{x}_i$ to the set of positive classes $\mathfrak{K}_{b,+} := \{k; \psi_{k,b} = +1\}$ as opposed to the set of negative classes $\mathfrak{K}_{b,-} := \{k; \psi_{k,b} = -1\}$ for task b. Thus, this calibrated membership probability can be regarded as a function of the membership probabilities $P(C = k|\vec{x}_i)$ involved in the current task:

$$\begin{aligned} p_{+,b}(\vec{x}_i) &:= P(C \in \mathfrak{K}_{b,+} | C \in \mathfrak{K}_{b,+} \cup \mathfrak{K}_{b,-}, \vec{x}_i) \\ &= \frac{\sum_{k \in \mathfrak{K}_{b,+}} P(C = k|\vec{x}_i)}{\sum_{k \in \mathfrak{K}_{b,+} \cup \mathfrak{K}_{b,-}} P(C = k|\vec{x}_i)}. \end{aligned} \quad (5.6)$$

The values $P(C = k|\vec{x}_i)$ solving Equation (5.6) would be the membership probabilities that reflect the assessment uncertainty. However, considering the additional constraint to membership probabilities

$$\sum_{k=1}^{K} P(C = k|\vec{x}_i) = 1 \quad (5.7)$$

there exist only $K - 1$ free parameters $P(C = k|\vec{x}_i)$, but at least K equations for the one–against rest approach and even more for all–pairs $(K(K-1)/2)$ or other ECOC approaches. Since the number of free parameters is always smaller than the number of constraints, no unique solution for the calculation of membership probabilities is possible and an approximative solution has to be found instead. Therefore, Zadrozny (2001) supplies a generalized version of the coupling algorithm of Hastie & Tibshirani (1998). This optimization method finds the estimated conditional probabilities $\hat{p}_{+,b}(\vec{x}_i)$ as realizations of a Binomial distributed random variable with an expected value $\mu_{b,i}$ in a way that

- $\hat{p}_{+,b}(\vec{x}_i)$ generate unique membership probabilities $\hat{P}(C = k|\vec{x}_i)$,
- $\hat{P}(C = k|\vec{x}_i)$ meet the probability constraint of Equation (5.7) and
- $\hat{p}_{+,b}(\vec{x}_i)$ have minimal Kullback–Leibler divergence to observed $p_{+,b}(\vec{x}_i)$.

The algorithm Hastie & Tibshirani (1998) and Zadrozny (2001) both use for their optimization is the model for paired comparisons by Bradley & Terry (1952), see the mentioned papers for details.

5.4. Calibration based on the Dirichlet distribution

The idea underlying the following multivariate calibration method is to transform the combined binary classification task outputs into realizations of a Dirichlet distributed random vector $\vec{P} \sim \mathcal{D}(h_1, \ldots, h_K)$ and regard the elements as membership probabilities $P_k := P(C = k|\vec{x})$.

Due to the concept of *Well–Calibration* by DeGroot & Fienberg (1983), see also Section 2.4.4, one wants to achieve that the confidence in the assignment to a particular class converges to the probability for this class. This requirement can be easily attained with a Dirichlet distributed random vector by choosing parameters h_k proportional to the a–priori probabilities $\pi_1, \ldots, \pi_K$ of classes, since elements P_k of the Dirichlet distributed random vector have expected values

$$\mathbf{E}(P_k) \;=\; h_k / \sum_{j=1}^{K} h_j \,, \tag{5.8}$$

see Johnson *et al.* (2002).

5.4.1. Dirichlet distribution

Initially, the *Dirichlet distribution* is presented in the definition of Johnson *et al.* (2002). A random vector $\vec{P} = (P_1, \ldots, P_K)'$ generated by K independently χ^2–

distributed random variables $S_k \sim \chi^2(2 \cdot h_k)$, see Section A.1.1, with

$$P_k = \frac{S_k}{\sum_{j=1}^{K} S_j} \quad (k = 1, 2, \ldots, K) \tag{5.9}$$

is Dirichlet distributed with parameters $h_1, \ldots, h_K$.

The probability density function of a random vector $\vec{P} = (P_1, \ldots, P_K)$ with parameters $h_1, \ldots, h_K$ distributed according to a Dirichlet distribution is

$$f(\vec{p}) = \frac{\Gamma\left(\sum_{j=1}^{K} h_j\right)}{\prod_{j=1}^{K} \Gamma(h_j)} \left(1 - \sum_{j=1}^{K-1} p_j\right)^{h_K - 1} \prod_{j=1}^{K-1} p_j^{h_j - 1}$$

with Gamma function

$$\Gamma(x) = \int_0^\infty u^{x-1} \exp(-u) \mathbf{d}u \,.$$

Let $S_1, \ldots, S_{K-1}$ be independent Beta distributed random variables where

$$S_k \sim \mathcal{B}(h_k, h_{k+1} + \ldots + h_K) \,.$$

Then $\vec{P} = (P_1, \ldots, P_K)' \sim \mathcal{D}(h_1, \ldots, h_K)$ where the P_k are defined by

$$P_k = S_k \cdot \prod_{j=1}^{k-1} (1 - S_j) \,, \tag{5.10}$$

see Aitchison (1963).

Furthermore, the single elements P_k of a Dirichlet distributed random vector are Beta distributed, since they are derived by a quotient of two χ^2– distributed random variables

$$S_k \sim \chi^2(2 \cdot h_k)$$
$$\sum_{j=1}^{K} S_j \sim \chi^2\left(\sum_{j=1}^{K} 2 \cdot h_j\right)$$

see Equation (5.9). As seen in Section A.1.2 Beta distributed random variables are derived similarly by a quotient of two χ^2–distributed variables. Hence, the values derived by Equation 5.10 are Beta distributed.

5.4.2. Dirichlet calibration

In the following the Dirichlet distribution is used as an essential for the introduced calibration method. Initially, instead of applying a univariate calibration method one simply normalizes the output vectors $s_{i,+1,b}$ generated by the classification rules to proportions with simple normalization (4.1). It is required to use a smoothing factor $\rho = 1.05$ in (4.1) so that $p_{i,+1,b} \in \,]0,1[$. Reason for this is the subsequent calculation of the product of associated binary proportions for each class $k \in \{1, \ldots, K\}$

$$r_{i,k} := \left[\prod_{b:\psi_{k,b}=+1} p_{i,+1,b} \cdot \prod_{b:\psi_{k,b}=-1} (1 - p_{i,+1,b}) \right] \tag{5.11}$$

analogous to the formula by Aitchison (1963), see Equation (5.10). Since the elements derived with Equation (5.10) are Beta distributed, the $r_{i,k}$ are regarded as realizations of a Beta distributed random variable $R_k \sim \mathcal{B}(\alpha_k, \beta_k)$. Therefore, parameters α_k and β_k are estimated from the training set by the method of moments as in Section 4.4.1.

To derive a multivariate Dirichlet distributed random vector, the $r_{i,k}$ can be transformed to realizations of a uniformly distributed random variable

$$u_{i,k} := F_{\mathcal{B},\hat{\alpha}_k,\hat{\beta}_k}(r_{i,k}) \ .$$

With application of the inverse of the χ^2–distribution function these uniformly distributed random variables are further transformed into χ^2–distributed random variables

$$F^{-1}_{\chi^2, h_k}(u_{i,k})$$

with h_k degrees of freedom.

Normalization is applied to yield the desired realizations of a Dirichlet distributed random vector $\vec{P} \sim \mathcal{D}(h_1, \ldots, h_K)$ with elements

$$\hat{p}_{i,k} := \frac{F^{-1}_{\chi^2, h_k}(u_{i,k})}{\sum_{j=1}^{K} F^{-1}_{\chi^2, h_j}(u_{i,j})}$$

The parameters $h_1, \ldots, h_K$ have to be set proportional to the class frequencies

$\pi_1, \ldots, \pi_K$ of the particular classes, since the expected value of the elements of a Dirichlet distributed vector is equal to the h_k, see Equation (5.8). Therefore, the $\hat{p}_{i,k}$ are realizations of a corresponding random variable which has an expected value that is equal to the correctness. Hence, membership probabilities generated with this calibration method are well–calibrated as in the sense of DeGroot & Fienberg (1983).

In the optimization procedure the factor $m \in \{1, 2, \ldots, 2 \cdot N\}$ with respective parameters $h_k = m \cdot \pi_k$ is chosen which leads to membership probabilities that score highest in terms of performance determined by the geometric mean of the three performance measures **CR** (2.3), $1 - $ **RMSE** (2.4) and **WCR** (2.5), see Chapter 2. The usage of the geometric mean is based on the idea of desirability indices, see Harrington (1965). This has got the advantage in leading to an overall value of 0 if one of the performance measures is equal to 0.

As in the univariate version of the Dirichlet distribution, the Beta distribution as defined by Garczarek (2002), the factor m is the analog to the dispersion parameter, see Section A.1.2. Therefore, the grid search is performed on the range $N \pm N$, i. e. with omitting 0 from 1 to $2 \cdot N$.

Alternatives for combination As an alternative to the derivation of the Beta distributed random variables with using the product of binary proportions, see Equation (5.11), several possibilities occur.

One option is to use the geometric mean instead

$$r_{i,k} := \left[\prod_{b:\psi_{k,b}=+1} p_{i,+1,b} \cdot \prod_{b:\psi_{k,b}=-1} (1 - p_{i,+1,b}) \right]^{\frac{1}{\#\{\psi_{k,b} \neq 0\}}} \tag{5.12}$$

with the formula by Aitchison (1963), see Equation (5.10), slightly changed from product to geometric mean so that values for different classes become more comparable.

The geometric mean is preferred to the arithmetic mean of proportions, since the product is well applicable for proportions, especially when they are skewed. Such skewed proportions are likely to occur when using the one–against rest approach in situations with high class numbers, since here the negative strongly outnumber the positive class observations.

Furthermore, it is reasonable to use a trimmed geometric mean instead, since especially with higher numbers of classes outliers are more likely. Using a robust estimate is a good method to handle outliers. To calculate the geometric trimmed mean one has to remove a truncation amount of γ percent of each sides of the ordered scores. Then one can use the same formula as in Equation (5.12). In the experimental analyses using a truncation of $\gamma = 10$ percent on both sides leads to the best results.

The experimental analysis in Section 6.2 will compare the Dirichlet calibration based on these three different approaches.

Multivariate application Some classification methods, e. g. the multivariate ANN, see Section 3.2, are able to directly generate multivariate probability matrices. Therefore, it is desirable to supply a direct calibration technique as in Figure 5.1 that re–calibrates the elements $p_{i,k}$ of a multivariate probability matrix.

These elements can be regarded as elements of a Dirichlet distributed random vector and hence as Beta distributed random variables. Thus, $r_{i,k}$ as in Equation (5.11) can be simply supplied by

$$r_{i,k} := p_{i,k}. \tag{5.13}$$

Afterwards, the parameters of the Beta distribution can be estimated as above by the method of moments and subsequently the Dirichlet calibration can be directly applied to such probability matrices for a re–calibration.

6. Analyzing uni– and multivariate calibrators

The following experimental analysis compares the calibration methods introduced in the previous chapters. At first, the univariate calibration procedures that were presented in Chapter 4 will be compared according to their performance in Section 6.1. Performance of a calibration method is quantified with the correctness rate (2.3) reflecting precision and the calibration measure (2.6) reflecting probabilistic reliability, see Section 2.4.1 and 2.4.4, respectively. 10–fold cross–validation is used to supply reliable generalized performance measures.

Secondly, the multivariate extensions of calibration methods, see Chapter 5, are compared in Section 6.2 according to their performance on various multi–class data sets.

The analysis is conducted with software R, see Ihaka & Gentleman (1996).

6.1. Experiments for two–class data sets

The currently known univariate calibration procedures were presented in Chapter 4. These calibrators will be compared in the following for the calibration of outputs from several classifiers, i. e. the regularization methods from Chapter 3 and other classifiers.

This includes a calibration of unnormalized scores generated by the regularization methods:

1. $L2$ Support Vector Machine ($L2$–SVM)

2. Artificial Neural Network (ANN).

Additionally, an analysis of *re–calibrating* membership probabilities generated by the following methods is presented:

1. Naive Bayes

2. Tree.

Exemplary for a tree algorithm the R procedure *tree* is used. Tree, see Chapter 7 in Ripley (1996), has got a similar tree learning and pruning algorithm as the tree procedure *CART* by Breiman et al. (1984).

The two latter classifiers produce membership values which claim to reflect the assessment uncertainty, but analyses by Zadrozny & Elkan (2001b) and Domingos & Pazzani (1996), respectively, show that these probabilities are inappropriate in reflect the assessment uncertainty.

The applied univariate calibration methods are the Assignment Value algorithm (av), Bayes calibration with using the asymmetric Laplace distribution (bay–alap), Isotonic (ir), Logistic (lr) and Piecewise Logistic Regression (plr). The calibrated probabilities are compared to scores generated by SVM or ANN calibrated with simple normalization and to the original probabilities generated by the Tree and Naive Bayes classifiers, respectively.

6.1.1. Two–class data sets

The following experiments are based on nine different data sets with binary class attributes. Most of the data sets which are presented in Table 6.1 origin from the UCI Repository of Machine Learning, see Newman et al. (1998).

All data sets only consist of numerical feature variables and do not contain any missing values, so that regularization methods are applicable.

The data sets are chosen so that every combination of a small, middle and high number of attributes and a small, middle and high number of observations is included once, see Table 6.2 for details.

Table 6.1.: Data sets with two classes – characteristics

Data set	N	p	Origin
Banknotes	200	5	Flury & Riedwyl (1983)
Breast–Cancer	683	10	Newman et al. (1998)
Fourclass	862	2	Ho & Kleinberg (1996)
German Credit	1000	24	Newman et al. (1998)
Haberman's Survival	306	3	Newman et al. (1998)
Heart	270	13	Newman et al. (1998)
Ionosphere	351	32	Newman et al. (1998)
Pima Indians Diabetes	768	8	Newman et al. (1998)
Sonar, Mines vs. Rocks	208	60	Newman et al. (1998)

Table 6.2.: Data sets with two classes – categorization

	p small ($p \leq 5$)	p middle ($5 < p \leq 20$)	p high ($p > 20$)
N small ($N \leq 300$)	*Banknotes* 200\5	*Heart* 270\13	*Sonar* 208\60
N middle ($300 < N \leq 700$)	*Haberman* 306\3	*Breast* 683\10	*Ionosphere* 351\32
N high ($N > 700$)	*Fourclass* 862\2	*Pima Indians* 768\8	*German* 1000\24

6.1.2. Results for calibrating classifier scores

In the following the calibration performance of the previously described univariate calibration procedures is analyzed. Therefore, the basis for this analysis are classifier scores generated by the regularization methods $L2$–SVM and ANN for the data sets from Table 6.1.

Calibration of $L2$–SVM classifier scores First of all, we analyze the calibration of unnormalized scores for the $L2$–SVM classifier.
Regarding the calibration of $L2$–SVM outputs, the calibration with assignment values is the most stable one, the performance measures for this method are shown in Table 6.3. This algorithm performs very well on the five data sets *Banknotes*, *Heart*, *Ionosphere*, *Pima Indians* and *Sonar*. Except *Pima Indians* where Bayes has slightly better **Cal**, this method yields best results for these data sets. Regarding the other four data sets this method yields also moderate, but not the best results. Compared to $L2$–SVM with simple normalization the AV–algorithm is always better, but for these four data sets it is outperformed by other methods. Especially for the *German* and the *Haberman* data set all other calibrators perform better. Anyway, this method yields for all data sets, except *Fourclass* and *Pima Indians*, a good **Cal**.
The major competitor for the AV–algorithm is the Logistic Regression (LR) which has good results for most of the data sets, too. Compared to the assignment value algorithm and to simple normalization this calibrator is better for the four data sets *Breast*, *German*, *Fourclass* and *Haberman*, for the latter two LR even yields the overall best performance. Furthermore, for *Pima Indians* and *Banknotes* LR has slightly poorer correctness, but compared to simple normalization strongly improved **Cal**. Anyway, for *Heart* and *Ionosphere* the results are quite poor and for *Sonar* very poor.
According to Zhang & Yang (2004) Piecewise Logistic Regression (PLR), see Section 4.2.2, gives an advantage to calibration with Logistic Regression. For calibrating $L2$–SVM classifier scores this assumption cannot be confirmed, since PLR is only better once, for data set *Sonar*. Logistic Regression is much better for *Banknotes*, *Ionosphere* and *Heart* and slightly better for the five remaining data sets.

Table 6.3.: Performance Measures for the $L2$–SVM classifier

	Banknotes		Breast		Fourclass	
	CR	Cal	CR	Cal	CR	Cal
$P_{L2\text{-SVM}}(C|\vec{x})$	0.990	0.630	0.954	0.627	0.761	0.506
$P_{\text{av}}(C|\vec{x})$	0.990	0.878	0.954	0.893	0.767	0.665
$P_{\text{bay-alap}}(C|\vec{x})$	0.965	0.941	0.931	0.853	0.773	0.673
$P_{\text{ir}}(C|\vec{x})$	0.930	0.872	0.963	0.877	0.773	0.677
$P_{\text{lr}}(C|\vec{x})$	0.985	0.911	0.960	0.903	0.787	0.704
$P_{\text{plr}}(C|\vec{x})$	0.940	0.902	0.959	0.902	0.784	0.686
	German Credit		Haberman		Heart	
	CR	Cal	CR	Cal	CR	Cal
$P_{L2\text{-SVM}}(C|\vec{x})$	0.649	0.408	0.679	0.423	0.829	0.461
$P_{\text{av}}(C|\vec{x})$	0.674	0.618	0.692	0.533	0.829	0.726
$P_{\text{bay-alap}}(C|\vec{x})$	0.700	0.607	0.722	0.386	0.811	0.724
$P_{\text{ir}}(C|\vec{x})$	0.710	0.577	0.722	0.378	0.811	0.728
$P_{\text{lr}}(C|\vec{x})$	0.700	0.613	0.722	0.408	0.814	0.733
$P_{\text{plr}}(C|\vec{x})$	0.700	0.607	0.722	0.390	0.807	0.725
	Ionosphere		Pima Indians		Sonar	
	CR	Cal	CR	Cal	CR	Cal
$P_{L2\text{-SVM}}(C|\vec{x})$	0.928	0.468	0.743	0.489	0.783	0.452
$P_{\text{av}}(C|\vec{x})$	0.928	0.718	0.743	0.631	0.788	0.633
$P_{\text{bay-alap}}(C|\vec{x})$	0.860	0.723	0.743	0.659	0.730	0.570
$P_{\text{ir}}(C|\vec{x})$	0.786	0.600	0.727	0.647	0.716	0.576
$P_{\text{lr}}(C|\vec{x})$	0.905	0.758	0.742	0.659	0.730	0.564
$P_{\text{plr}}(C|\vec{x})$	0.831	0.679	0.738	0.659	0.745	0.582

Isotonic Regression (IR) delivers high variation in performance. On the one hand, this method works fine on *Haberman* and is even best in precision on *German* and *Breast*, though with poorer **Cal**. On the other hand, this method performs poorly on *Heart* and it has even the worst performance for *Banknotes*, *Ionosphere*, *Pima Indians* and *Sonar*.

The Bayes algorithm by Bennett (2002) yields mediocre results. This method has competitive performance for *Fourclass*, *German*, *Haberman* and *Pima Indians*, but the performance for the rest, especially *Breast*, *Ionosphere* and *Sonar*, is quite poor. For the first two data sets this occurs due to the small number of misclassified observations. Hence, the distributional parameters for the misclassified scores cannot be estimated adequately.

Concluding the analysis for calibrating two–class $L2$–SVM scores, the Assignment Value algorithm and the calibration with Logistic Regression yield good and overall more or less robust results. The three other calibrators are not recommendable due to a high variation in performance.

Calibration of ANN–classifier scores The results for the analysis of the calibration performance on the second classifier – an Artificial Neural Network –to an is given in Table 6.4.

Compared to the original ANN output the AV–algorithm yields equal precision for all data sets which indicates that the initial assignment to classes is not changed by this calibrator. This method only changes the relation between probabilities which is not necessarily a bad case, see Section 2.4.4. Additionally, the **Cal** is better for seven out of the nine data sets. The exceptions are *Sonar* and *Fourclass* for which it is slightly and clearly worse, respectively.

Comparing the **Cal** of the Assignment Value Algorithm and Logistic Regression, AV has got better values for *Banknotes* and *Haberman* while LR has for *German*, *Heart* and *Ionosphere*. For the four remaining data sets the values are quite similar. In the analysis occurs a little bit more variation in precision for LR than for AV. Anyway, this is more often a decrease, since only for *German* LR has a slightly better correctness rate, but it is even clearly worse for *Pima Indians* and *Sonar*.

Comparing Piecewise (PLR) to Standard Logistic Regression (LR), PLR is better for *Haberman* and *Pima Indians*, also slightly for *Banknotes* and *Fourclass*.

Table 6.4.: Performance Measures for the ANN–classifier

	Banknotes		Breast		Fourclass	
	CR	**Cal**	**CR**	**Cal**	**CR**	**Cal**
$P_{\text{ANN}}(C|\vec{x})$	0.995	0.899	0.966	0.701	0.829	0.189
$P_{\text{av}}(C|\vec{x})$	0.995	0.982	0.966	0.939	0.829	0.083
$P_{\text{bay-alap}}(C|\vec{x})$	0.995	0.990	0.967	0.933	0.752	0.126
$P_{\text{ir}}(C|\vec{x})$	0.990	0.980	0.963	0.936	0.837	0.069
$P_{\text{lr}}(C|\vec{x})$	0.995	0.971	0.964	0.939	0.825	0.080
$P_{\text{plr}}(C|\vec{x})$	0.995	0.974	0.959	0.936	0.827	0.077
	German Credit		Haberman		Heart	
	CR	**Cal**	**CR**	**Cal**	**CR**	**Cal**
$P_{\text{ANN}}(C|\vec{x})$	0.753	0.561	0.718	0.556	0.796	0.661
$P_{\text{av}}(C|\vec{x})$	0.753	0.600	0.718	0.569	0.796	0.676
$P_{\text{bay-alap}}(C|\vec{x})$	0.752	0.602	0.705	0.543	0.792	0.659
$P_{\text{ir}}(C|\vec{x})$	0.755	0.617	0.712	0.557	0.792	0.681
$P_{\text{lr}}(C|\vec{x})$	0.755	0.623	0.712	0.560	0.796	0.684
$P_{\text{plr}}(C|\vec{x})$	0.751	0.618	0.722	0.575	0.777	0.675
	Ionosphere		Pima Indians		Sonar	
	CR	**Cal**	**CR**	**Cal**	**CR**	**Cal**
$P_{\text{ANN}}(C|\vec{x})$	0.900	0.700	0.695	0.561	0.793	0.675
$P_{\text{av}}(C|\vec{x})$	0.900	0.829	0.695	0.584	0.793	0.669
$P_{\text{bay-alap}}(C|\vec{x})$	0.871	0.798	0.671	0.558	0.798	0.658
$P_{\text{ir}}(C|\vec{x})$	0.903	0.827	0.709	0.600	0.769	0.602
$P_{\text{lr}}(C|\vec{x})$	0.900	0.843	0.688	0.583	0.783	0.668
$P_{\text{plr}}(C|\vec{x})$	0.857	0.799	0.699	0.598	0.701	0.549

Overall, this method has even the best results for *Haberman* and second best for *Pima Indians*. However, there occur more and higher negative deviations, since the results are poor for *Breast* and *Heart*, even very poor for *Ionosphere* and *Sonar*. For these four data sets and also for *German Credit* PLR has got the worst results. Therefore, this calibrator is not recommendable for an application to ANN classifier scores.

As for *L2*–SVM, see above, the behavior of Isotonic Regression (IR) has its pros and cons. This method yields best performance measures for *Fourclass*, *Ionosphere* and *Pima Indians*. For the first two, the correctness is even clearly better than for all other calibrators. Also for *German Credit* this method has got the highest correctness. However, the **CR** for all the other five data sets is worse than **CR** for simple normalization and the AV–algorithm. Except the *Haberman* data set **CR** for IR is always worse than for LR, too.

For calibrating ANN scores the Bayes method yields results with high variation. On the one hand, this calibrator has got the best performance for *Breast*, *Sonar* and *Banknotes*. Though, the performance does not differ very much between all calibrators for the latter data set. On the other hand, the performance is the worst one for *Haberman*, *Pima Indians* and *Fourclass* with being extremely poor for the latter. Since the performance is also quite bad for *Ionosphere*, this method can not be recommended for calibrating ANN scores in spite to the good outcomes for the first three mentioned data sets.

Summary of calibration results Concluding the results for calibrating regularization classifier scores for two–class data sets, it is to note that the Assignment Value algorithm gives the best and most robust results. Compared to initial classifier scores this method produces membership probabilities that deliver a performance increase for almost all data sets. The major competitor is the Logistic Regression calibration by Platt (1999) which is better for some data sets, but overall it yields compared to AV more often worse than better results.

The Piecewise Logistic Regression calibrator yields no improvement in performance compared to standard LR and is therefore not a recommendable alternative. The two remaining calibrators, Isotonic Regression and the Bayes approach have got their highs and lows. These methods change the whole structure of classifier scores and hence a higher variation in performance is not surprising.

Anyway, since in most cases this variation does not lead to an increase in performance, these calibrators cannot be recommended for calibrating classifier scores generated by regularization methods. Comparing the three latter methods with the AV–algorithm and the Logistic Regression it occurs that these methods adhere an intrinsic higher complexity and are therefore more likely to be object of overfitting, see Section 2.4.2.

6.1.3. Results for re–calibrating membership probabilities

In the following we will regard the re–calibration of membership probabilities generated by the Naive Bayes (NB) classifier and a Tree procedure. These classification methods have issues with a reflection of the assessment uncertainty, see Section 5.1.

Re–Calibration of NB–classifier scores One of the methods which produces insufficient membership probabilities is the Naive Bayes classifier for which the results are presented in Table 6.5.

In regarding the re–calibration of Naive Bayes classifier scores for two–class data sets the Assignment Value (AV) algorithm again yields good results. For the data sets *Banknotes*, *German Heart*, *Ionosphere* and *Pima Indians* this method has overall the best results. Compared to NB scores the AV–algorithm is also performing better on *Fourclass*, although most of the other methods yield higher values here. Regarding *Haberman* and *Breast* the Assignment Value calibration results are slightly poorer than Naive Bayes while only for *Sonar* the performance of the AV–algorithm is inadequate.

In re–calibrating NB scores, Logistic Regression (LR) is again the major competitor for the AV–algorithm, although it yields only better results for data sets *Haberman* and *Sonar*. Correctness is similar for *Banknotes*, *Breast* and *Heart*, but the calibration measure is clearly worse in these cases. Compared to NB scores the performance of LR is also good on *German* and *Ionosphere*, but all other calibrators are better for the first data set, while for the latter LR is only outperformed by the AV–algorithm. Considering *Fourclass* and *Pima Indians* LR performs slightly poorer than most of the other methods.

Table 6.5.: Performance Measures for NB–classifier

	Banknotes		Breast		Fourclass	
	CR	**Cal**	**CR**	**Cal**	**CR**	**Cal**
$P_{\text{NB}}(C|\vec{x})$	0.995	0.979	0.963	0.919	0.757	0.678
$P_{\text{av}}(C|\vec{x})$	0.995	0.979	0.960	0.915	0.769	0.674
$P_{\text{bay–alap}}(C|\vec{x})$	0.500	0.375	0.961	0.916	0.781	0.679
$P_{\text{ir}}(C|\vec{x})$	0.500	0.375	0.781	0.576	0.783	0.715
$P_{\text{lr}}(C|\vec{x})$	0.995	0.837	0.961	0.801	0.754	0.676
$P_{\text{plr}}(C|\vec{x})$	0.605	0.416	0.963	0.722	0.793	0.719
	German Credit		Haberman		Heart	
	CR	**Cal**	**CR**	**Cal**	**CR**	**Cal**
$P_{\text{NB}}(C|\vec{x})$	0.719	0.568	0.741	0.550	0.840	0.741
$P_{\text{av}}(C|\vec{x})$	0.738	0.648	0.735	0.573	0.844	0.758
$P_{\text{bay–alap}}(C|\vec{x})$	0.730	0.639	0.741	0.547	0.837	0.746
$P_{\text{ir}}(C|\vec{x})$	0.736	0.637	0.751	0.581	0.837	0.667
$P_{\text{lr}}(C|\vec{x})$	0.728	0.628	0.745	0.620	0.840	0.664
$P_{\text{plr}}(C|\vec{x})$	0.736	0.635	0.732	0.575	0.844	0.649
	Ionosphere		Pima Indians		Sonar	
	CR	**Cal**	**CR**	**Cal**	**CR**	**Cal**
$P_{\text{NB}}(C|\vec{x})$	0.823	0.685	0.761	0.642	0.677	0.505
$P_{\text{av}}(C|\vec{x})$	0.888	0.798	0.765	0.667	0.663	0.576
$P_{\text{bay–alap}}(C|\vec{x})$	0.840	0.730	0.714	0.643	0.701	0.560
$P_{\text{ir}}(C|\vec{x})$	0.641	0.537	0.763	0.659	0.533	0.512
$P_{\text{lr}}(C|\vec{x})$	0.854	0.641	0.755	0.652	0.692	0.553
$P_{\text{plr}}(C|\vec{x})$	0.641	0.573	0.759	0.654	0.543	0.518

In contrast to calibration of scores generated by regularization methods, the Bayes approach performs comparable to the other two methods for a re–calibration on Naive Bayes scores. For *Sonar* this method is even the best and for *Fourclass* it is better than the previously mentioned calibrators and the Naive Bayes classifier. Additionally, for *Breast, German, Haberman, Ionosphere* the performance is good though the Bayes calibrator is not the overall best here. Nevertheless, for *Heart* it is slightly outperformed by most of the other methods while for *Pima Indians* it is clearly poorer than all other calibrators. Very poor results occur only for data set *Banknotes*. The almost perfect classification in combination with the hence justified extreme Naive Bayes probabilities leads to the contradicting fact, that the Bayes approach is not adequate here. With the lack of variation in probabilities and the lack of wrong assignment the parameters for the Asymmetric Laplace distribution cannot be estimated correctly, since the distribution has to be also estimated for misclassified scores, see Section 4.3.

In contrast to the three previous methods, Piecewise Logistic and Isotonic Regression again deliver no increase compared to uncalibrated classifier scores. As the Bayes approach, these methods cannot handle the probabilities which are almost all extreme and therefor lack variation which were generated by NB for *Banknotes*. Furthermore, these methods perform miserable for *Ionosphere* and *Sonar*. While Isotonic Regression is in comparison also poor for *Breast* and *Heart*, PLR has even the best **CR** for these two data sets. With the remaining four data sets PLR again shows high variation in performance in ranging from overall best *(Fourclass)* over good *(German)* and mediocre *(Pima Indians)* to poor *(Haberman)*.

Apart from the five previously mentioned data sets which the Isotonic Regression is performing very poorly for, it is the best calibrator for *Haberman*, almost best and only outperformed by PLR for *Fourclass* and also good for *German* and *Pima Indians*.

Re–Calibration of Tree classifier scores Table 6.6 shows performance results for the re–calibration of classifier scores generated with the R procedure *tree*.

As for the previously presented classifiers the Assignment Value (AV) algorithm is also the best method for calibrating Tree classifier scores, although the performance is not much better than the original scores. In comparison, the performance

Table 6.6.: Performance Measures for Tree–classifier

	Banknotes		Breast		Fourclass	
	CR	Cal	CR	Cal	CR	Cal
$P_{\text{TREE}}(C\|\vec{x})$	0.985	0.902	0.935	0.901	0.964	0.941
$P_{\text{av}}(C\|\vec{x})$	0.985	0.924	0.938	0.905	0.967	0.945
$P_{\text{bay-alap}}(C\|\vec{x})$	0.500	0.381	0.918	0.860	0.954	0.917
$P_{\text{ir}}(C\|\vec{x})$	0.920	0.847	0.379	0.386	0.961	0.854
$P_{\text{lr}}(C\|\vec{x})$	0.960	0.894	0.869	0.547	0.946	0.880
$P_{\text{plr}}(C\|\vec{x})$	0.540	0.543	0.434	0.379	0.965	0.889
	German Credit		Haberman		Heart	
	CR	Cal	CR	Cal	CR	Cal
$P_{\text{TREE}}(C\|\vec{x})$	0.665	0.520	0.683	0.560	0.707	0.617
$P_{\text{av}}(C\|\vec{x})$	0.659	0.514	0.679	0.554	0.714	0.619
$P_{\text{bay-alap}}(C\|\vec{x})$	0.597	0.502	0.647	0.524	0.692	0.570
$P_{\text{ir}}(C\|\vec{x})$	0.667	0.414	0.712	0.568	0.685	0.592
$P_{\text{lr}}(C\|\vec{x})$	0.667	0.507	0.689	0.531	0.659	0.555
$P_{\text{plr}}(C\|\vec{x})$	0.679	0.513	0.673	0.554	0.688	0.597
	Ionosphere		Pima Indians		Sonar	
	CR	Cal	CR	Cal	CR	Cal
$P_{\text{TREE}}(C\|\vec{x})$	0.891	0.829	0.707	0.592	0.701	0.553
$P_{\text{av}}(C\|\vec{x})$	0.891	0.827	0.712	0.595	0.701	0.551
$P_{\text{bay-alap}}(C\|\vec{x})$	0.871	0.762	0.710	0.570	0.721	0.556
$P_{\text{ir}}(C\|\vec{x})$	0.888	0.830	0.696	0.583	0.716	0.590
$P_{\text{lr}}(C\|\vec{x})$	0.866	0.808	0.709	0.577	0.725	0.580
$P_{\text{plr}}(C\|\vec{x})$	0.894	0.843	0.701	0.590	0.663	0.554

is equal or similar for almost all data sets. It is slightly better for *Breast*, *Fourclass* and *Pima Indians*, (almost) equal for *Banknotes*, *Ionosphere* and *Sonar* as well as slightly worse for *German Credit* and *Haberman*. Only for *Heart* the difference between correctness rates is a bit higher, but the AV–algorithm is still not far better here.

In contrast to previous calibration results Logistic Regression (LR) is less competitive in calibrating tree scores. There are only two data sets – *Haberman* and *Sonar* – where the method delivers an advantage to original estimates with being even the best method for the latter one. Additionally, this method is similar to original scores for *Pima Indians* and *German*. Anyway, for the remaining five data sets LR yields no improvement. While LR has got a correctness value for *Banknotes* which can be still regarded as high, correctness is anyway much better for the AV–algorithm and the original tree scores. Furthermore, it is even the worst method for *Fourclass*, *Heart* and *Ionosphere*. For *Breast* the descent in performance to original tree scores is definite. The pruning procedure in the tree algorithm leads for these data sets to scores with a small number of parameter values which is shown in Figure A.2 in the appendix. Hence, the linear function of the log odds cannot be estimated appropriately here. In contrast to that, the linear function can be estimated more accurately for the Naive Bayes scores, since this method leads to scores with more variation, see Figure A.1. Summing this up, since LR is only good for two data sets here, but worse than original scores for five data sets calibration with Logistic Regression cannot be recommended for calibration if scores lack variation.

Piecewise Logistic Regression (PLR), the Bayes approach and Isotonic Regression (IR) deliver again no improvement to original classifier scores. The performance of PLR is unacceptable for *Banknotes* and *Breast*, for the first one Bayes is also very poor and Isotonic Regression for the latter one. PLR is indeed the best method for *German Credit* as well as for *Ionosphere* and has also comparable performance measures for *Fourclass*. Nevertheless, for the four remaining data sets PLR leads to a clear decrease in correctness and is therefore not recommendable.

Bayes is only very good for *Sonar* and good for *Pima Indians*, while IR is the best calibrator for *Haberman*, second best for *Sonar* and quite comparable for *Ionosphere*. Results for the corresponding remaining data sets are poor for correctness and/or calibration measure. Hence, these calibrators cannot be recommended here again.

Summary of re–calibration results In conclusion, calibration of two-class classifier scores generated by tree algorithms cannot be regarded as necessary. Even the advantage of the best–performing AV–algorithm is in doubt, since measures are almost always comparable. All other calibration methods do not yield an improvement but a decrease in performance compared to the tree scores.
Considering calibration of Naive Bayes scores the AV–algorithm, Logistic Regression and also the Bayes approach yield good results.

6.2. Experiments for multi–class data sets

Subsequent to the analysis of calibrating classifier scores for two–class data sets, in this section the performance of the multivariate extensions of calibration procedures is compared. The different multivariate calibrators were presented in Chapter 5.
The correctness and calibration measures, derived with 10–fold cross–validation, will be presented for each of the following combinations of classification method and reduction approach:

1. $L2$–SVM classifier with one–against rest reduction,

2. $L2$–SVM classifier with all–pairs reduction,

3. ANN classifier with one–against rest reduction,

4. ANN classifier with all–pairs reduction.

Both ANN reduction approaches are compared to a direct multivariate ANN.
Performance analyses for the Naive Bayes classifier are not included for several reasons. This classifier is directly applicable for multi–class data sets and the analyses, see Section 6.1.3, only show an increase in performance by applying calibration for some data sets. Last but not least the better approach in using the Naive Bayes classifier is to initially check the assumption, i. e. independence

of feature variables, and then decide whether this method is applied to this particular data set or not.

Furthermore, performance analyses for the tree algorithm are omitted here, since results for two classes, see Table 6.6, did not show a performance increase induced by any calibration.

Anyway, for each of these combinations the following multivariate calibration procedures are applied:

1. Original simply normalized scores combined with coupling,

2. Assignment value calibrated scores combined with coupling (av),

3. Scores calibrated by Logistic Regression combined with coupling (lr),

4. Dirichlet calibration with using the geometric mean (diri–g–mean),

5. Dirichlet calibration with the trimmed geometric mean (diri–g–trim),

6. Dirichlet calibration with using the product (diri–prod),

7. direct Dirichlet calibration (diri–direct).

The last calibration method is only applied to the direct multivariate ANN.
Univariate calibration methods using the Bayes approach including the asymmetric Laplace distribution (bay–alap), Isotonic (ir) and Piecewise Logistic Regression (plr) did not yield acceptable results in two–class calibration, see Section 6.1. Therefore and to keep comparison manageable, these methods are omitted in the following analysis. For the sake of completeness results these calibrators are incorporated in Tables A.1 to A.4 in the appendix.

6.2.1. K–class data sets

Nine different data sets with multi–class labels, i. e. $K > 2$, were chosen for the following analyses of multivariate calibrators. Like the selected two–class data sets the multi–class ones have got only numerical feature variables and no missing values.

Most of the data sets origin from the UCI Repository Of Machine Learning, see

Newman et al. (1998), see the overview in Table 6.7.

As another indication for further analyses class distributions in these data sets are presented with absolute numbers and in percent in Tables 6.8 and 6.9.

The Tables 6.8 and 6.9 separate the data sets by the relative amount of individual classes into balanced and unbalanced class distribution, respectively.

Table 6.7.: Data sets with K classes

	N	p	K	Origin
B3	157	13	4	Heilemann & Münch (1996)
Balance	625	4	3	Newman et al. (1998)
Ecoli	336	7	8	Newman et al. (1998)
Glass	214	9	6	Newman et al. (1998)
Iris	150	4	3	Newman et al. (1998)
Segment	2310	19	7	Newman et al. (1998)
Vehicle	846	18	4	Newman et al. (1998)
Wine	178	13	3	Newman et al. (1998)
Zoo	101	16	7	Newman et al. (1998)

Table 6.8.: Class distribution of K–class data sets – unbalanced class distribution

Balance	Class	B	L	R				
	N	49	288	288				
	%	8	46	46				
Glass	Class	1	2	3	5	6	7	
	N	70	76	17	13	9	29	
	%	33	36	8	6	4	14	
Ecoli	Class	cp	im	imL	imS	imU	om	omL	pp
	N	143	77	2	2	35	20	5	52
	%	43	23	1	1	10	6	1	15
Zoo	Class	1	2	3	4	5	6	7	
	N	41	20	5	13	4	8	10	
	%	41	20	5	13	4	8	10	

Table 6.9.: Class distribution of K–class data sets – balanced class distribution

$B3$	Class	1	2	3	4		
	N	59	24	47	27		
	%	38	15	30	17		
$Iris$	Class	setosa	versicolor	virginica			
	N	50	50	50			
	%	33	33	33			
$Segment$	Class	brickface	cement	foliage	grass	path	sky	window
	N	330	330	330	330	330	330	330
	%	14	14	14	14	14	14	14
$Vehicle$	Class	bus	opel	saab	van		
	N	218	212	217	199		
	%	26	25	26	24		
$Wine$	Class	1	2	3			
	N	59	71	48			
	%	33	40	27			

6.2.2. Calibrating $L2$–SVM scores in multi–class situations

In the following the results for calibrating $L2$–SVM classification outputs for multi–class data sets are presented for the one–against rest and the all–pairs approach separately.

$L2$–SVM classification with one–against rest reduction Table 6.10 presents correctness and calibration measures for $L2$–SVM classification with one–against rest reduction.

Considering one–against rest $L2$–SVM classification results differ very much between data sets. Especially for the $Glass$ data set the performance is very poor which indicates that the grid search on parameters, see Section 3.1.3, is not sufficient here.

Nevertheless, calibration procedures differ in their performance, too. The Assign-

Table 6.10.: Performance for the $L2$–SVM using one–against rest reduction

	B3		Balance		Ecoli	
	CR	Cal	CR	Cal	CR	Cal
$P_{L2\text{-SVM}}(C\|\vec{x})$	0.713	0.569	0.892	0.551	0.833	0.727
$P_{\text{av}}(C\|\vec{x})$	0.694	0.590	0.894	0.754	0.791	0.784
$P_{\text{lr}}(C\|\vec{x})$	0.687	0.558	0.892	0.831	0.824	0.736
$P_{\text{diri-g-mean}}(C\|\vec{x})$	0.694	0.708	0.892	0.795	0.839	0.856
$P_{\text{diri-g-trim}}(C\|\vec{x})$	0.694	0.708	0.892	0.795	0.839	0.856
$P_{\text{diri-prod}}(C\|\vec{x})$	0.675	0.679	0.894	0.796	0.827	0.830
	Glass		Iris		Segment	
	CR	Cal	CR	Cal	CR	Cal
$P_{L2\text{-SVM}}(C\|\vec{x})$	0.481	0.625	0.926	0.560	0.838	0.666
$P_{\text{av}}(C\|\vec{x})$	0.453	0.632	0.893	0.739	0.822	0.755
$P_{\text{lr}}(C\|\vec{x})$	0.476	0.583	0.920	0.673	0.796	0.723
$P_{\text{diri-g-mean}}(C\|\vec{x})$	0.476	0.647	0.913	0.815	0.857	0.870
$P_{\text{diri-g-trim}}(C\|\vec{x})$	0.476	0.647	0.913	0.815	0.857	0.870
$P_{\text{diri-prod}}(C\|\vec{x})$	0.467	0.677	0.920	0.790	0.858	0.861
	Vehicle		Wine		Zoo	
	CR	Cal	CR	Cal	CR	Cal
$P_{L2\text{-SVM}}(C\|\vec{x})$	0.686	0.563	0.966	0.524	0.910	0.737
$P_{\text{av}}(C\|\vec{x})$	0.695	0.635	0.966	0.786	0.891	0.831
$P_{\text{lr}}(C\|\vec{x})$	0.667	0.581	0.966	0.710	0.910	0.791
$P_{\text{diri-g-mean}}(C\|\vec{x})$	0.700	0.717	0.971	0.850	0.920	0.878
$P_{\text{diri-g-trim}}(C\|\vec{x})$	0.700	0.717	0.971	0.850	0.920	0.878
$P_{\text{diri-prod}}(C\|\vec{x})$	0.703	0.706	0.971	0.811	0.891	0.892

ment Value algorithm (AV) is not as robust as for two–class tasks, see Section 6.1.2. While the calibration measure is better than coupled normalized $L2$–SVM scores for all data sets, the correctness is poorer for six data sets *(B3, Ecoli, Glass, Iris, Segment* and *Zoo)* and comparable or slightly better only for the three remaining data sets *Balance, Vehicle* and *Wine*.

Variation in performance of Logistic Regression (LR) can be called similar to AV, but overall LR has got slight advantages, since it is clearly better for four data sets *(Ecoli, Glass, Iris* and *Zoo)*, but AV only for *Segment* and *Vehicle*.

Anyway, compared to $L2$–SVM LR is similar in **CR** and with better **Cal** for four data sets *(Balance, Iris, Wine* and *Zoo)*. The remaining data sets range from slightly worse *(Ecoli* and *Glass)* over quite poor *(B3* and *Vehicle)* to even poorer for *Segment*.

Since Dirichlet calibration with geometric mean does not differ in performance to calibration with the trimmed geometric mean, description of results will be presented bundled here. Compared to original scores these calibrators are better for the five data sets *Ecoli, Segment, Vehicle, Wine* and *Zoo*. For further data sets performance ranges from similar for *Balance*, over slightly worse *(Glass* and *Iris)* to clearly worse *(B3)*. On (almost) all data sets **Cal** is better for this method than for $L2$–SVM scores and the two other calibrators.

Calibration performance is almost as good for the Dirichlet calibration with product. Regarding correctness this calibration method has got some more variation. Dirichlet calibration with product is the most precise method for four data sets *(Balance, Segment, Vehicle* and *Wine)* and has also a good **CR** for *Iris*. Except *Balance* all these data sets have a balanced class distribution. Anyway, precision is mediocre for the unbalanced data set *Ecoli* and is even the worst for *B3* as well as for the two unbalanced data sets *Glass* and *Zoo*.

Summing these findings up, Dirichlet calibration with the geometric (trimmed) mean appears to be the most robust calibration method in combination with one–against rest $L2$–SVM classification. Although, Dirichlet calibration using the product has got best performance for three data sets and can be recommended for data sets with a balanced class distribution, the other Dirichlet calibrator is preferable due to its robustness. Further reason for the better performance of the geometric (trimmed) mean is that the unbalanced class distributions in the binary learning are even intensified by the one–against rest approach. Additionally, it is to note that $L2$–SVM without calibration is actually the best method for three data sets.

$L2$–SVM classification with all–pairs reduction Table 6.11 presents the correctness and calibration measures for the calibration of scores generated by the $L2$–SVM using all-pairs reduction.

The first point to mention here is that the original $L2$–SVM scores yield best results for four data sets *(B3, Vehicle, Wine* and *Zoo)*, while the Assignment Value algorithm (AV) is the best method for two other data sets *(Ecoli* and *Glass)*. This calibrator is also good for three further data sets *(Balance, Iris* and *Segment)*. While the AV–algorithm is worse than $L2$–SVM for the four remaining data sets, it is still the (one of the) best calibrators here.

Logistic Regression (LR) has got more variation in its results than the AV–algorithm. It ranges from best method *(Balance* and *Iris)* over mediocre *(Ecoli, Glass* and *Segment)* and poor *(B3, Wine* and *Zoo)* to worst *(Vehicle)*.

Performance of Dirichlet calibration with geometric trimmed mean is always better or equal to performance for applying the geometric mean. It is the best overall method for *Segment* and the best calibrator for two other data sets *(B3* and *Zoo)*. Additionally, performance ranges from comparable *(Balance)* over mediocre *(Glass, Iris, Vehicle* and *Wine)* to quite poor for *Ecoli*. The latter data set has got a highly unbalanced class distribution with several classes that have few observations, i. e. only two to five observations belong to three ("imL","imS","omL") out of the eight classes, see Table 6.8.

Finally, the Dirichlet calibration with product does not yield comparable results for calibrating $L2$–SVM using all-pairs. This method is unacceptable poor for *Ecoli* which has a highly unbalanced class distribution, see above. It is also the worst for two unbalanced *(Glass* and *Zoo)*, but also for three balanced data sets *(B3, Segment* and *Wine)*. Furthermore, it is poor for *Iris* and only comparable for the remaining two data sets *(Balance* and *Vehicle)*.

Summing up the findings for $L2$–SVM with all–pairs reduction, the AV algorithm is the best calibration method here, but original scores are better for four data sets. The Dirichlet calibration with geometric trimmed mean is an alternative for data sets with a balanced or moderately unbalanced class distribution, since this method is only poor for *Ecoli*.

Table 6.11.: Performance Measures for the $L2$–SVM using all–pairs reduction

	B3		Balance		Ecoli	
	CR	Cal	CR	Cal	CR	Cal
$P_{L2\text{-SVM}}(C|\vec{x})$	0.738	0.531	0.873	0.514	0.824	0.646
$P_{\text{av}}(C|\vec{x})$	0.694	0.672	0.889	0.781	0.842	0.808
$P_{\text{lr}}(C|\vec{x})$	0.668	0.652	0.889	0.853	0.815	0.799
$P_{\text{diri-g-mean}}(C|\vec{x})$	0.694	0.696	0.888	0.741	0.794	0.724
$P_{\text{diri-g-trim}}(C|\vec{x})$	0.694	0.696	0.888	0.741	0.794	0.724
$P_{\text{diri-prod}}(C|\vec{x})$	0.636	0.617	0.886	0.837	0.488	0.725
	Glass		Iris		Segment	
	CR	Cal	CR	Cal	CR	Cal
$P_{L2\text{-SVM}}(C|\vec{x})$	0.528	0.599	0.893	0.524	0.926	0.603
$P_{\text{av}}(C|\vec{x})$	0.630	0.686	0.94	0.734	0.935	0.833
$P_{\text{lr}}(C|\vec{x})$	0.551	0.674	0.966	0.726	0.933	0.879
$P_{\text{diri-g-mean}}(C|\vec{x})$	0.570	0.676	0.880	0.690	0.940	0.822
$P_{\text{diri-g-trim}}(C|\vec{x})$	0.570	0.676	0.920	0.707	0.940	0.822
$P_{\text{diri-prod}}(C|\vec{x})$	0.476	0.662	0.893	0.847	0.888	0.843
	Vehicle		Wine		Zoo	
	CR	Cal	CR	Cal	CR	Cal
$P_{L2\text{-SVM}}(C|\vec{x})$	0.703	0.561	0.971	0.476	0.960	0.622
$P_{\text{av}}(C|\vec{x})$	0.699	0.676	0.960	0.787	0.940	0.808
$P_{\text{lr}}(C|\vec{x})$	0.634	0.660	0.938	0.749	0.920	0.740
$P_{\text{diri-g-mean}}(C|\vec{x})$	0.691	0.673	0.949	0.745	0.950	0.825
$P_{\text{diri-g-trim}}(C|\vec{x})$	0.691	0.673	0.949	0.745	0.950	0.825
$P_{\text{diri-prod}}(C|\vec{x})$	0.692	0.696	0.926	0.850	0.861	0.907

Summary of calibrating $L2$–SVM classifier scores Concluding the analyses on calibrating $L2$–SVM scores, the Dirichlet calibration with the geometric (trimmed) mean is the best method for calibrating one–against rest outcomes and the AV–algorithm is the best calibrator for all–pairs outcomes. Comparing these two methods the AV–algorithm is better for four data sets *(Glass, Iris, Segment and Zoo)*, while results are quite similar for four *(B3, Balance, Ecoli and Vehicle)* out of the remaining five data sets.

Furthermore, the Dirichlet calibration with geometric trimmed mean can be recommended for calibrating all–pairs $L2$–SVM scores, if data sets have a balanced or moderately unbalanced class distribution.

Finally, the Bayes approach (bay–alap), Isotonic (IR) and Piecewise Logistic Regression PLR yield results with high variation, see Tables A.1 and A.2 in the appendix. As in the analyses of two–class data sets in Section 6.1, these methods cannot be recommended in calibration of multi–class $L2$–SVM outcomes.

6.2.3. Calibrating ANN–classification outcomes for $K > 2$

In the following performance measures will be presented for calibrating Artificial Neural Networks (ANN) with using the one–against rest and the all–pairs reduction.

In contrast to the $L2$–SVM above, the reduction approaches are also compared in the following analyses to a multivariate ANN as described in Section 3.2.

ANN–classification with one–against rest reduction Table 6.12 shows correctness and calibration measures for calibration of scores generated by ANN–classification with using the one–against rest reduction method.

The Assignment Value (AV) algorithm yields quite good results for calibrating one–against rest ANN–scores, although with some variation. AV is once the best calibrator *(Glass)* and has also very high performance measures for four further data sets *(Balance, Segment, Wine and Zoo)*. For the remaining data sets it ranges from comparable *(B3, Vehicle and Ecoli)* to worst in correctness *(Iris)*.

Logistic Regression (LR) is better than AV with beating this method six times. LR is the best method for two data sets *(Ecoli and Segment)* as well as very good

Table 6.12.: Performance for ANN–classification with one–against rest reduction

	B3		Balance		Ecoli	
	CR	**Cal**	**CR**	**Cal**	**CR**	**Cal**
$P_{\text{ANN-direct}}(C\|\vec{x})$	0.700	0.727	0.982	0.893	0.860	0.845
$P_{\text{ANN-rest}}(C\|\vec{x})$	0.662	0.684	0.953	0.638	0.869	0.794
$P_{\text{av}}(C\|\vec{x})$	0.662	0.683	0.956	0.895	0.863	0.868
$P_{\text{lr}}(C\|\vec{x})$	0.675	0.696	0.958	0.898	0.869	0.874
$P_{\text{diri-g-mean}}(C\|\vec{x})$	0.643	0.663	0.918	0.853	0.866	0.875
$P_{\text{diri-g-trim}}(C\|\vec{x})$	0.643	0.663	0.918	0.853	0.866	0.875
$P_{\text{diri-prod}}(C\|\vec{x})$	0.662	0.687	0.926	0.876	0.526	0.713
$P_{\text{diri-direct}}(C\|\vec{x})$	0.700	0.719	0.982	0.748	0.854	0.814
	Glass		Iris		Segment	
	CR	**Cal**	**CR**	**Cal**	**CR**	**Cal**
$P_{\text{ANN-direct}}(C\|\vec{x})$	0.598	0.680	0.966	0.881	0.654	0.758
$P_{\text{ANN-rest}}(C\|\vec{x})$	0.649	0.678	0.973	0.810	0.909	0.725
$P_{\text{av}}(C\|\vec{x})$	0.658	0.710	0.966	0.948	0.923	0.896
$P_{\text{lr}}(C\|\vec{x})$	0.635	0.709	0.973	0.940	0.927	0.900
$P_{\text{diri-g-mean}}(C\|\vec{x})$	0.649	0.709	0.973	0.958	0.909	0.934
$P_{\text{diri-g-trim}}(C\|\vec{x})$	0.649	0.709	0.973	0.958	0.909	0.934
$P_{\text{diri-prod}}(C\|\vec{x})$	0.579	0.692	0.973	0.934	0.912	0.934
$P_{\text{diri-direct}}(C\|\vec{x})$	0.574	0.661	0.966	0.836	0.637	0.754
	Vehicle		Wine		Zoo	
	CR	**Cal**	**CR**	**Cal**	**CR**	**Cal**
$P_{\text{ANN-direct}}(C\|\vec{x})$	0.695	0.708	0.887	0.760	0.950	0.868
$P_{\text{ANN-rest}}(C\|\vec{x})$	0.743	0.632	0.960	0.788	0.970	0.898
$P_{\text{av}}(C\|\vec{x})$	0.737	0.745	0.966	0.932	0.960	0.941
$P_{\text{lr}}(C\|\vec{x})$	0.732	0.751	0.955	0.927	0.960	0.908
$P_{\text{diri-g-mean}}(C\|\vec{x})$	0.737	0.744	0.966	0.949	0.950	0.949
$P_{\text{diri-g-trim}}(C\|\vec{x})$	0.737	0.744	0.966	0.949	0.950	0.949
$P_{\text{diri-prod}}(C\|\vec{x})$	0.739	0.741	0.960	0.929	0.960	0.814
$P_{\text{diri-direct}}(C\|\vec{x})$	0.679	0.702	0.893	0.845	0.950	0.908

for four other ones *(B3, Balance, Iris* and *Zoo)*. Nevertheless, it is worse than most of the other calibrators for the three remaining data sets *(Glass, Vehicle* and *Wine)*, although the differences are only small here.

Dirichlet calibration with geometric (trimmed) mean yields results which compared to other calibrators vary more between data sets. On the one hand, this calibrator is (almost) best in Well–Calibration for five data sets *(Ecoli, Glass, Iris, Vehicle* and *Wine)* with (very) good precision and has still high performance measures for three of the four remaining data sets *(Balance, Segment* and *Zoo)*. On the other hand, performance is poorer than most of the other calibrators for *Segment and Zoo*, while it is even worst for the two remaining data sets *(B3* and *Balance)*.

Dirichlet calibration with product again is an alternative for balanced data sets and not at all for data sets with an unbalanced class distribution. This method is the best calibrator once *(Vehicle)* and two times *(Iris* and *Zoo)* one of the calibrators with the highest correctness. It is comparable for three data sets *(Wine, Segment* and *B3)*, although sometimes outperformed by another method here. For the unbalanced data sets *Balance* and *Glass* the performance is poorer for Dirichlet calibration with product than for (most of) the other calibrators. Finally, for *Ecoli* it is even unacceptable low due to the highly unbalanced class distribution for *Ecoli* in which three of the eight classes hold only one percent of the observations, see Table 6.8.

Direct Dirichlet calibration is the best method for two data sets *(B3* and *Balance)* and comparable for *Ecoli*, but is the worst calibrator for six data set *(Glass, Iris, Segment, Vehicle, Wine* and *Zoo)*. These worse results are due to the fact that the multivariate direct ANN yields poorer results here than the one–against rest ANN.

Concluding the results for ANN–classification with one–against rest reduction, all one–against rest calibrators can be recommended, except Dirichlet calibration with product for data sets with unbalanced class distributions. Logistic Regression is the best method for four data sets and beats the Assignment Value algorithm six times, but AV yields good results, too. Dirichlet calibration with geometric (trimmed) mean is the best calibrator considering Well–Calibration, since it has the highest **Cal** for five data sets *(Ecoli, Iris, Segment, Wine* and *Zoo)* and also comparable for two further ones *(Glass* and *Vehicle)*. Finally, direct application of ANN with or without calibration yields poorer results in most cases.

ANN–classification with all–pairs reduction Table 6.13 presents correctness and calibrations measures for calibrating ANN scores based on an all–pairs reduction.

In calibrating all–pairs ANN–scores with the Assignment Value algorithm (AV) performance is good for almost all data sets. The only exception is the *B3* data set where this method is clearly worse than original scores and Dirichlet calibration with the product. For four data sets *(Glass, Iris, Vehicle* and *Wine)* AV has comparable correctness with good to very good **Cal**. For two of the four remaining data sets *(Ecoli* and *Segment)* AV is even the best method, while for *Zoo* it is only outperformed by the Dirichlet methods and for *Balance* only by the direct methods.

As in calibrating one–against rest ANN–scores Logistic Regression (LR) yields very good results for application to all–pairs outcomes, too. For six data sets *(Ecoli, Glass, Iris, Segment, Vehicle* and *Wine)* correctness and calibration measure range from best to third and are also very good for *Balance*. Nevertheless, Logistic Regression one of the worst methods for *B3* and worse than the Dirichlet calibrators for *Zoo*, although still with a high correctness here.

The Dirichlet calibration with geometric (trimmed) mean is not recommendable for calibrating scores generated by an ANN with using all–pairs on unbalanced data sets. Although it has the highest correctness for the moderately unbalanced data set *Zoo*, it is the worst method for three data sets *(Balance, Ecoli* and *Glass)* where classes are not balanced. While being also the worst method for *B3*, it is comparable to the other calibrators for the four remaining balanced data sets *(Iris, Segment, Vehicle* and *Wine)*.

As above Dirichlet calibration with product is recommendable for data sets with balanced to moderately unbalanced class distributions. Performance is the best for three balanced data sets *(B3, Iris* and *Wine)*. For the two remaining balanced data sets *(Vehicle* and *Segment)* this method is worse than LR and AV, but it has got still high performance measures for the latter. With one exception *(Vehicle)* Dirichlet calibration with product has got (almost) the best **Cal** for the balanced data sets. For the two data sets with moderately unbalanced class distributions *(Glass* and *Zoo)* it is comparable to others. Finally, for the two highly unbalanced

Table 6.13.: Performance Measures for the ANN using all–pairs reduction

	B3		Balance		Ecoli	
	CR	Cal	CR	Cal	CR	Cal
$P_{\text{ANN–direct}}(C|\vec{x})$	0.700	0.727	0.982	0.893	0.860	0.845
$P_{\text{ANN–pairs}}(C|\vec{x})$	0.719	0.700	0.910	0.558	0.848	0.705
$P_{\text{av}}(C|\vec{x})$	0.707	0.736	0.974	0.932	0.877	0.884
$P_{\text{lr}}(C|\vec{x})$	0.700	0.729	0.972	0.937	0.863	0.853
$P_{\text{diri–g–mean}}(C|\vec{x})$	0.700	0.695	0.902	0.741	0.514	0.650
$P_{\text{diri–g–trim}}(C|\vec{x})$	0.700	0.695	0.902	0.741	0.514	0.650
$P_{\text{diri–prod}}(C|\vec{x})$	0.726	0.729	0.945	0.889	0.836	0.885
$P_{\text{diri–direct}}(C|\vec{x})$	0.700	0.719	0.982	0.748	0.854	0.814
	Glass		Iris		Segment	
	CR	Cal	CR	Cal	CR	Cal
$P_{\text{ANN–direct}}(C|\vec{x})$	0.598	0.680	0.966	0.881	0.654	0.758
$P_{\text{ANN–pairs}}(C|\vec{x})$	0.686	0.659	0.973	0.780	0.916	0.701
$P_{\text{av}}(C|\vec{x})$	0.686	0.733	0.973	0.945	0.950	0.942
$P_{\text{lr}}(C|\vec{x})$	0.691	0.720	0.973	0.907	0.950	0.940
$P_{\text{diri–g–mean}}(C|\vec{x})$	0.630	0.700	0.973	0.817	0.942	0.801
$P_{\text{diri–g–trim}}(C|\vec{x})$	0.630	0.700	0.973	0.817	0.942	0.801
$P_{\text{diri–prod}}(C|\vec{x})$	0.677	0.730	0.973	0.941	0.918	0.938
$P_{\text{diri–direct}}(C|\vec{x})$	0.574	0.661	0.966	0.836	0.637	0.754
	Vehicle		Wine		Zoo	
	CR	Cal	CR	Cal	CR	Cal
$P_{\text{ANN–direct}}(C|\vec{x})$	0.695	0.708	0.887	0.760	0.950	0.868
$P_{\text{ANN–pairs}}(C|\vec{x})$	0.757	0.585	0.943	0.781	0.910	0.786
$P_{\text{av}}(C|\vec{x})$	0.755	0.769	0.943	0.929	0.930	0.919
$P_{\text{lr}}(C|\vec{x})$	0.756	0.770	0.943	0.920	0.930	0.764
$P_{\text{diri–g–mean}}(C|\vec{x})$	0.754	0.759	0.943	0.826	0.960	0.719
$P_{\text{diri–g–trim}}(C|\vec{x})$	0.754	0.759	0.943	0.826	0.960	0.719
$P_{\text{diri–prod}}(C|\vec{x})$	0.749	0.727	0.943	0.923	0.940	0.919
$P_{\text{diri–direct}}(C|\vec{x})$	0.679	0.702	0.893	0.845	0.950	0.908

data sets *(Balance* and *Ecoli)* it is clearly worse than AV and LR, although still with high correctness for *Balance*.

Direct Dirichlet calibration is performing best for one data set *(Balance)* and comparable to good for three further data sets *(Ecoli , Iris* and *Zoo)* with similar results for the direct Multivariate ANN here. Also similar is the poor performance for four further data sets *(B3, Glass, Vehicle* and *Wine)*. For the remaining data set *(Segment)* the direct multivariate ANN and hence Dirichlet calibration is highly outperformed by the all–pairs methods.

Summarizing the results for calibrating all–pairs ANN scores, Logistic Regression and the Assignment Value algorithm have got high performance. The Dirichlet calibration with product is competitive for data sets with a balanced or moderately unbalanced class distribution, while Dirichlet calibration with geometric (trimmed) mean is only recommendable for data sets with a balanced class distribution. Direct Multivariate calibration has got the same variation as the direct Multivariate ANN with being outperformed by all–pairs methods more often than vice versa.

Summary for calibrating multi–class ANN–scores The Assignment Value algorithm and Logistic Regression can be recommended for calibrating multi–class Artificial Neural Network scores. These methods yield high results for both reduction approaches on almost all data sets.

Due to comparable good results the Dirichlet calibrators can be recommended for balanced data sets, but are not appropriate if class distributions are (highly) unbalanced. In calibrating one–against rest ANN–scores the geometric trimmed mean is preferable while for calibrating all–pairs the application of the product yields more robust results.

Similar to the analyses of two–class data sets in Section 6.1 the results for the Isotonic (IR) and Piecewise Logistic Regression (PLR) are object to high variation, see Tables A.3 and A.4 in the appendix. Therefore, these methods cannot be recommended in calibration of multi–class ANN outcomes. Exception is the Bayes approach (bay–alap) which performs comparable to other calibrators for most of the data sets.

Comparing the reduction approaches to the direct ANN with or without calibration both reduction approaches yield more robust results.

7. Conclusion and Outlook

This thesis delivers a framework for the generation of membership probabilities in polychotomous classification tasks with supervised learning. Machine Learning and Statistical Classification are the two research fields that drive innovation and further development of classification methods. As a consequence, membership values that are generated by classifiers can be separated into two groups – unnormalized scores and membership probabilities. When comparing these two types of membership values, membership probabilities have the advantage that they reflect the assessment uncertainty. Therefore, these probabilities are the preferable outcome, especially in situations where the classification outcome is subject to post–processing. If a classification method only generates unnormalized scores, calibration can be used to supply membership probabilities.

Regularization is a major example for a type of classifier that does not produce membership probabilities but unnormalized scores. Therefore, the two most common regularization methods – Artificial Neural Networks (ANN) and Support Vector Machine (SVM) – are the main classification procedures which generate outcomes that need to be calibrated. In application of calibration to SVM scores, it should be considered that analyses by Zhang (2004) show that it is preferable to use the $L2$–SVM instead of the $L1$–SVM, since scores generated by the $L2$–SVM reflect more information on estimating membership probabilities, see Section 3.3. While the correctness rate (**CR**) is the essential goodness criterion for measuring the performance of a classification method, a further measure is necessary to indicate the quality of the assessment uncertainty reflection. Therefore, the *Well–Calibration Ratio* (**WCR**) based on the concept by DeGroot & Fienberg (1983) is introduced in Section 2.4.4. Using the geometric mean of **WCR** and **RMSE** which measures the effectiveness in assignment to classes, goodness of calibration can be quantified by a single value.

Moreover, it is possible that classifiers produce membership values that do not reflect the assessment uncertainty sufficiently and therefore need to be *re–calibrated*. According to analyses from Domingos & Pazzani (1996) and Zadrozny & Elkan (2001b) too extreme and hence inappropriate membership probabilities are generated by both the Naive Bayes classifier and Tree procedures, respectively. In contrast to that, analyses in Section 6.1 show that re–calibrated membership probabilities do not produce an increase in the calibration measure compared to membership probabilities derived by the R procedure *tree*, although they do for Naive Bayes membership probabilities. Still, for the Naive Bayes classifier it is nonetheless recommended to check its assumptions and to decide afterwards whether to apply this method or not.

Due to the above, this thesis focusses on calibrating scores generated by regularization methods such as the SVM or an ANN. Particularly, the SVM classifier has got an additional drawback, since it is only directly applicable to two–class problems.

Hence, calibration techniques are usually introduced for dichotomous classification tasks. Chapter 4 presents currently known *univariate* calibration techniques, i. e. the Assignment Value algorithm, Logistic and Piecewise Logistic Regression, the Bayes approach with an asymmetric Laplace distribution as well as Isotonic Regression. These methods estimate membership probabilities for one class and for the other class the membership probabilities is estimated with the complement.

The experimental analysis of univariate calibration methods in Section 6.1 shows that the Assignment Value algorithm by Garczarek (2002) performs most robust on various data sets and different classification techniques. The major competitor for the AV–algorithm is the Logistic Regression by Platt (1999) which yields comparable results in calibrating ANN, SVM and Naive Bayes classifier scores. Although calibrating tree scores does not seem to be necessary as for the other mentioned classifiers, it has to be noted that this method does not perform as well as for the other classifiers described.

The Bayes approach using an asymmetric Laplace distribution as well as the Isotonic and Piecewise Logistic Regression are not robust enough to supply appropriate membership probabilities over various different data sets. These methods are more likely to overfit the classifier outcomes due to their higher complexity.

An exception to the Naive Bayes classifier is the application of the Bayes method which was designed by Bennett (2003) specially for calibrating Naive Bayes scores and yields competitive results here.

Regarding Classification tasks with number of classes K greater than two regularization methods usually apply a reduction to several binary problems followed by a combination of the evaluated binary membership values to one membership probability for each class. The two common algorithms for a reduction to binary classification tasks are the one–against rest and the all-pairs approaches. While the one–against rest method is based on a smaller number of learning procedures than the latter one, the all–pairs binary learning rules are based on fewer observations and therefore are faster to learn. Another advantage of the all-pairs reduction algorithm is that positive and negative classes are balanced in the binary learning. In contrast, the one–against rest reduction classes are unbalanced, since one class is considered to be positive and all remaining classes negative.

Allwein *et al.* (2000) generalize these two reduction approaches with the idea of using *ECOC–matrices* and further present two additional approaches for the reduction of polychotomous classification tasks, the sparse and the complete ECOC–matrix. Calculations in Section 5.2 show that the reduction method using a sparse ECOC–matrix is not appropriate for classification tasks with less than nine classes, since the proposed derivation of such a kind of ECOC–matrix is not sufficiently elaborated. The size of a complete ECOC–matrix is exponentially increasing with the number of classes. Hence, the reduction algorithm is not practicable for data sets with more than five classes.

The common approach for combining the membership values supplied for the binary learning procedures all–pairs and one–against rest is the pairwise coupling algorithm by Hastie & Tibshirani (1998). After learning the rules for the binary tasks with a subsequent univariate calibration in the first step, multi–class membership probabilities are generated by a pairwise coupling of these calibrated membership probabilities as the second step.

This thesis introduces the Dirichlet calibration as an alternative one–step multivariate calibrator which is applicable to binary outcomes as well as applicable to multivariate probability matrices. This method combines the simply normalized binary classification scores or the columns of the probability matrix with either the geometric trimmed mean or the product and uses these proportions to generate a Dirichlet distributed random vector. By choosing the parameters for the Dirichlet distribution proportional to a–priori probabilities these membership probabilities are usually well–calibrated, see analyses in Section 6.2. Precision is comparable to the application of pairwise coupling and univariate calibration with either Assignment Value algorithm or Logistic Regression for data sets with a balanced or moderately unbalanced class distribution. Here, Dirichlet calibration with the geometric trimmed mean performs better with using $L2$–SVM and ANN with one–against rest while using the product yields better results in combination with an ANN and the all–pairs approach. With data sets where observations are highly unbalanced between classes, these calibrators occasionally yield poor performance. Especially, this occurs in situations where classes have got very low numbers of observations. Here, the parameters of the Beta distribution cannot be estimated correctly which can lead to a miscaculation of the Dirichlet distributed random vector. Therefore, Dirichlet calibration can only be recommended for an application on data sets with a balanced or lightly unbalanced distribution of classes. In these situations, the performance of this method is comparable to common calibration approaches but are easier to generate due to the one step algorithm.

With the introduction of the Dirichlet calibration method, this thesis supplies a calibrator which is a well-performing alternative for the derivation of membership probabilities in multi–class situations applicable on dichotomous and polychotomous classifier outcomes.

A more thorough analysis of performance in classifying K–class data sets could be supplied with further simulations on bigger data sets, e. g. *Letter*, *Dorothea*, *Spambase*, etc., see Newman *et al.* (1998). An option to investigate the decreasing performance of the Dirichlet calibration method on unbalanced data sets would be to simulate data sets with differing class distributions ranging from balanced over slightly unbalanced to highly unbalanced. Balanced and unbalanced class distributions could be generated with the uniform and the χ^2–distribution, respectively. Additionally, a calibration of membership values generated by further machine learning methods for classification like Boosting by Freund & Schapire (1995) or Random Forests by Breiman (2001) has to be evaluated.

An option to increase the performance of the Dirichlet calibration method on unbalanced data is the inclusion of a weighting procedure in the estimation of the

Dirichlet distributed random vector. If estimated elements of the random vector are in doubt for classes with a low number of observations these elements could be weighed low while classes with a reasonable number of observations have to be weighed high.

Bibliography

Aitchison, J. (1963): Inverse distributions and independent gamma distributed products of random variables. *Biometrika*, 50, 505–508.

Allwein, E. L., Schapire, R. E., & Singer, Y. (2000): Reducing multiclasss to binary: A unifying approach for margin classifiers. *Journal of Machine Learning Research*, 1, 113–141.

Anderson, T. W. (1958): *An Introduction to Multivariate Statistical Analysis*. New York: John Wiley & Sons.

Aronszajn, N. (1950): Theory of reproducing kernels. *Transactions of the American Mathematical Society*, 68, 337–404.

Bennett, P. N. (2002): Using asymmetric distributions to improve text classifier probability estimates: A comparison of new and standard parametric methods. *Technical Report CMU-CS-02-126*, Carnegie Mellon, School of Computer Science.

Bennett, P. N. (2003): Using asymmetric distributions to improve text classifier probability estimates. In: *Proceedings of the 26th Annual International ACM SIGIR Conference on Research and Development in Information Retrieval*, 111–118. Toronto: ACM Press.

Berger, J. O. (1985): *Statistical Decision Theory and Bayesian Analysis*. New York: Springer, 2nd edition.

Bradley, R. A. & Terry, M. (1952): The rank analysis of incomplete block designs, i: The method of paired comparisons. *Biometrika*, 39, 324–345.

Breiman, L. (2001): Random forests. *Machine Learning*, 45 (1), 5–32.

Breiman, L., Friedman, J. H., Olshen, R. A., & Stone, C. J. (1984): *Classification and regression trees*. Belmont: Wadsworth International Group.

Bridle, J. S. (1989): Probabilistic interpretation of feedforward classification network outputs, with relationships to statistical pattern recocginition. In: F. Fougelman-Soulie & J. Hérault (eds.) *Neuro–Computing: algorithms, architectures and applications (NATO ASI Series F: Computer and Systems Science, volume 68)*, 227–236. Les Arcs: Springer.

Brier, G. W. (1950): Verfication of forecasts expressed in terms of probability. *Monthly Weather Review*, 78, 1–3.

Burges, C. J. C. (1998): A tutorial on support vector machines for pattern recognition. *Data Mining and Knowledge Discovery*, 2 (2), 121–167.

Cortes, C. & Vapnik, V. (1995): Support-vector networks. *Machine Learning*, 20 (3), 273–297.

Courant, R. & Hilbert, D. (1953): *Methods of Mathematical Physics*. New York: John Wiley & Sons.

Crammer, K. (2000): On the learnability and design of output codes for multiclass problems. In: *In Proceedings of the Thirteenth Annual Conference on Computational Learning Theory*, 35–46.

Cristianini, N. & Shawe-Taylor, J. (2000): *An Introduction to Support Vector Machines and other kernel–based learning methods*. Cambridge: Cambridge University Press.

Cybenko, G. (1989): Approximations by superpositions of sigmoidal functions. *Mathematics of Controls, Signal and Systems*, 2, 303–314.

DeGroot, M. H. & Fienberg, S. E. (1983): The comparison and evaluation of forecasters. *The Statistician*, 32, 12–22.

Devroye, L. (1986): *Non-Uniform Random Variate Generation*. New York: Springer.

Dietterich, T. G. & Bakiri, G. (1995): Solving multiclasss learning problems via error–correcting output codes. *Journal of Artificial Intelligence Research*, 2, 263–286.

Domingos, P. & Pazzani, M. J. (1996): Beyond independence: Conditions for the optimality of the simple bayesian classifier. In: *Proceedings of the 13th International Conference on Machine Learning*, 105–112. Bari: Morgan Kaufmann Publishers.

Duan, K. & Keerthi, S. S. (2005): Which is the best multiclass svm method? an empirical study. In: *Proceedings of the Sixth International Workshop on Multiple Classifier Systems*, 278–285.

Duda, R. O., Hart, P. E., & Stork, D. G. (1973): *Pattern Classification and Scene Analysis*. New York: John Wiley & Sons.

Fisher, R. A. (1936): The use of multiple measurements in taxonomic problems. *Annals of Eugenics*, 7, 179–188.

Flury, B. N. & Riedwyl, H. (1983): *Angewandte multivariate Statistik*. Stuttgart: Gustav Fischer Verlag.

Freund, Y. & Schapire, R. (1995): A decision-theoretic generalization of on-line learning and an application to boosting. *Journal of Computer and System Sciences*, 55 (1), 119–139.

Friedman, J., Hastie, T., & Tibshirani, R. (2000): Additive logistic regression: a statistical view of boosting. *Annals of Statistics*, 28, 2000.

Garczarek, U. M. (2002): *Classification Rules in Standardized Partition Spaces*. Dissertation, Universität Dortmund. URL http://eldorado.uni-dortmund.de:8080/FB5/ls7/forschung/2002/Garczarek.

Garczarek, U. M. & Weihs, C. (2003): Standardizing the comparison of partitions. *Computational Statistics*, 18 (1), 143–162.

Hand, D. J. (1997): *Construction and Assessment of Classification Rules*. Chichester: John Wiley & Sons.

Harrington, E. C. (1965): The desirability function. *Indistrial Quality Control*, 21 (10), 494–498.

Hartung, J., Elpelt, B., & Klösener, K.-H. (2005): *Statistik: Lehr- und Handbuch der angewandten Statistik*. Oldenbourg, 14th edition.

Hastie, T. & Tibshirani, R. (1998): Classification by pairwise coupling. In: M. I. Jordan, M. J. Kearns, & S. A. Solla (eds.) *Advances in Neural Information Processing Systems*, volume 10. Cambridge: MIT Press.

Hastie, T., Tibshirani, R., & Friedman, J. H. (2001): *The Elements of Statistical Learning - Data Mining, Inference and Prediction*. New York: Springer.

Heilemann, U. & Münch, H. (1996): West german business cycles 1963–1994: A multivariate discriminant analysis. In: *CIRET–Conference in Singapore, CIRET–Studien 50*.

Ho, T. K. & Kleinberg, E. M. (1996): Building projectable classifiers of arbitrary complexity. In: *Proceedings of the 13th International Conference on Pattern Recognition*, 880–885. Wien.

Hosmer, D. W. & Lemeshow, S. (2000): *Applied logistic regression (Wiley Series in probability and statistics)*. Wiley-Interscience Publication.

Hsu, C.-W., Chang, C.-C., & Lin, C.-J. (2003): A practical guide to support vector classification. URL http://www.csie.ntu.edu.tw/~cjlin/papers/guide/guide.pdf.

Ihaka, R. & Gentleman, R. (1996): R: A language for data analysis and graphics. *Journal of Computational and Graphical Statistics*, 5 (3), 299–314.

Johnson, N. L., Kotz, S., & Balakrishnan, N. (1995): *Continuous Uniivariate Distributions*. New York: John Wiley & Sons, 2nd edition.

Johnson, N. L., Kotz, S., & Balakrishnan, N. (2002): *Continuous Multivariate Distributions*, volume 1, Models and Applications. New York: John Wiley & Sons, 2nd edition.

Kearns, M., Li, M., Pitt, L., & Valiant, L. G. (1987): On the learnability of boolean formulae. In: *Proceedings of the 19th annual ACM Symposium on Theory of computing*, 285–295. New York: ACM Press.

Kuhn, H. W. & Tucker, A. (1951): Nonlinear programming. In: *Proceedings of 2nd Berkeley Symposium on Mathematical Statistics and Probabilistics*, 481–492. Berkeley: University of California Press.

Lee, Y., Lee, Y., Lin, Y., Lin, Y., Wahba, G., & Wahba, G. (2004): Multicategory support vector machines, theory, and application to the classification of microarray data and satellite radiance data. *Journal of the American Statistical Association*, 99, 67–81.

Lehmann, E. L. (1983): *Theory of Point Estimation*. New York: Springer.

Lin, Y. (1999): Support vector machines and the bayes rule in classification. *Technical Report 1014*, Department of Statistics, University of Wisconsin, Madison.

Lugosi, G. & Vayatis, N. (2004): On the bayes-risk consistency of regularized boosting methods. *Annals of Statistics*, 32 (1), 30–55.

McCulloch, W. S. & Pitts, W. H. (1943): A logical calculus of the ideas immanent in nervous activity. *Bulletin of Mathematical Biophysics*, 32, 115–133.

Minsky, M. L. & Papert, S. A. (1969): *Perceptrons*. Cambridge: MIT Press.

Newman, D. J., Hettich, S., Blake, C. L., & Merz, C. J. (1998): UCI repository of machine learning databases. URL http://www.ics.uci.edu/~mlearn/MLRepository.html.

Platt, J. C. (1999): Probabilistic outputs for support vector machines and comparisons to regularized likelihood methods. In: A. J. Smola, P. Bartlett, B. Schölkopf, & D. Schuurmans (eds.) *Advances in Large Margin Classiers*, 61–74. Cambridge: MIT Press.

Quinlan, J. R. (1993): *C4.5 – Programs for Machine Learning*. San Mateo: Morgan Kaufmann Publishers.

Ripley, B. D. (1996): *Pattern Recognition and Neural Networks*. Cambridge: Cambridge University Press.

Rosenblatt, F. (1958): The perceptron: A probabilistic model for information storage and organization in the brain. *Psychological Review*, 65, 386–408.

Rosenblatt, F. (1962): *Principles of Neurodynamics: Perceptrons and the Theory of Brain Mechanisms*. New York: Spartan.

Rumelhart, D. E., Hinton, G. E., & Williams, R. J. (1986): Learning representations by error propagation. In: D. E. Rumelhart, J. L. McClelland, & the PDP Research Group (eds.) *Parallel Distributed Processing*, 318–362. Cambridge: MIT Press.

Schapire, R. E., Freund, Y., Bartlett, P., & Lee, W. S. (1998): Boosting the margin: a new explanation for the effectiveness of voting methods. *The Annals of Statistics*, 26, 322–330.

Schölkopf, B., Burges, C. J. C., & Vapnik, V. N. (1995): Extracting support data for a given task. In: U. M. Fayyad & R. Uthurusamy (eds.) *Proceedings of the First International Conference on Knowledge Discovery and Data Mining*. Menlo Park: AAAI Press.

Schölkopf, B., Sung, K.-K., Burges, C. J., Girosi, F., Niyogi, P., Poggio, T., & Vapnik, V. N. (1997): Comparing support vector machines with gaussian kernels to radial basis function classifiers. *IEEE Transactions on Signal Processing*, 45 (11), 2758–2765.

Shawe-Taylor, J. & Cristianini, N. (2004): *Kernel methods for pattern analysis*. Cambridge: Cambridge University Press.

Steinwart, I. (2005): Consistency of support vector machines and other regularized kernel classifiers. *IEEE Transactions on Information Theory*, 51 (1), 128–142.

Suykens, J. A. K., Lukas, L., van Dooren, P., de Moor, B. L. R., & Vandewalle, J. P. L. (1999): Least squares support vector machine classifiers: a large scale algorithm. In: *Proceedings of the European Conference on Circuit Theory and Design (ECCTD'99)*, 839–842. Stresa: Levrotto & Bella.

Suykens, J. A. K. & Vandewalle, J. P. L. (1999): Least squares support vector machine classifiers. *Neural Processing Letters*, 9 (3), 293–300.

Vapnik, V. N. (2000): *The Nature of Statistical Learning Theory*. New York: Springer.

Vogtländer, K. & Weihs, C. (2000): Business cycle prediction using support vector methods. *Technical Report 21/00*, SFB 475, University of Dortmund.

Wahba, G. (1998): Support vector machines, reproducing kernel hilbert spaces and the randomized gacv. In: A. J. Smola, C. J. C. Burges, & B. Schölkopf (eds.) *Advances in Kernel Methods: Support Vector Learning*, 69–87. Cambridge: MIT Press.

Weiss, S. M. & Kulikowski, C. A. (1991): *Computer systems that learn: classification and prediction methods from statistics, neural nets, machine learning, and expert systems*. San Mateo: Morgan Kaufmann Publishers.

Weston, J. & Watkins, C. (1998): Multi-class support vector machines. *Technical Report CSD-TR-98-04*, Royal Holloway University of London.

Zadrozny, B. (2001): Reducing multiclass to binary by coupling probability estimates. In: *Advances in Neural Information Processing Systems 14*. Vancouver: MIT Press.

Zadrozny, B. & Elkan, C. (2001a): Learning and making decisions when costs and probabilities are both unknown. In: *Proceedings of the Seventh International Conference on Knowledge Discovery and Data Mining*, 204–213. San Francisco: ACM Press.

Zadrozny, B. & Elkan, C. (2001b): Obtaining calibrated probability estimates from decision trees and naive bayes classifiers. In: *Proceedings of the Eighteenth International Conference on Machine Learning*, 609–616. San Francisco: Morgan Kaufmann Publishers.

Zadrozny, B. & Elkan, C. (2002): Transforming classifier scores into accurate multiclass probability estimates. In: *Proceedings of the Eighth International Conference on Knowledge Discovery and Data Mining*, 694–699. Edmonton: ACM Press.

Zhang, J. & Yang, Y. (2004): Probabilistic score estimation with piecewise logistic regression. In: *Proceedings of the 21st International Conference on Machine Learning*. Banff: ACM Press.

Zhang, T. (2004): Statistical behavior and consistency of classification methods based on convex risk minimization. *Annals of Statistics*, 32 (1), 56–85.

A. Appendix

A.1. Distributions

A.1.1. Gamma distribution

Necessary for the derivation of the Beta distribution is the Gamma Distribution, see Johnson et al. (1995). A random variable X is distributed according to $\mathcal{G}(b,p)$, if it has the probability density function

$$f_{\mathcal{G}}(x|b,p) \;=\; \frac{b^p}{\Gamma(p)} x^{p-1} \exp(-bx)$$

with Gamma function

$$\Gamma(p) \;=\; \int_0^\infty u^{p-1} \exp(-u) \mathbf{d} u\,.$$

A speciality of the Gamma distribution with parameters $b = 1/2$ and $p = \nu/2$ is the χ^2–distribution $\chi^2(\nu)$ with ν degrees of freedom.

A.1.2. Beta distribution

The quotient $X := \frac{G}{G+H}$ of two χ^2–distributed variables $G \sim \chi^2(2 \cdot \alpha)$ and $H \sim \chi^2(2 \cdot \beta)$ is Beta distributed according to $\mathcal{B}(\alpha, \beta)$ with parameters α and β. A Beta distributed random variable has got the density function

$$f_{\mathcal{B}}(x|\alpha,\beta) \;=\; \frac{1}{B(\alpha,\beta)} x^{\alpha-1}(1-x)^{\beta-1}$$

with Beta function

$$B(\alpha,\beta) \;=\; \frac{\Gamma(\alpha)\Gamma(\beta)}{\Gamma(\alpha+\beta)} \;=\; \int_0^1 u^{\alpha-1}(1-u)^{\beta-1} \mathbf{d} u\,.$$

Without loss of generality, Garczarek (2002) uses for better application of her calibration method a different parameterization for the Beta distribution in terms of expected value

$$p := E(x|\alpha, \beta) = \frac{\alpha}{\alpha + \beta}$$

and *dispersion* parameter

$$N := \alpha + \beta.$$

Calling N dispersion parameter is motivated by the fact that for fixed p this parameter determines the variance of a Beta distributed random variable

$$\text{Var}(x|\alpha, \beta) := \frac{\alpha\beta}{(\alpha+\beta)^2(\alpha+\beta+1)} = \frac{p(1-p)}{N-1}.$$

A.2. Further analyses of calibration results

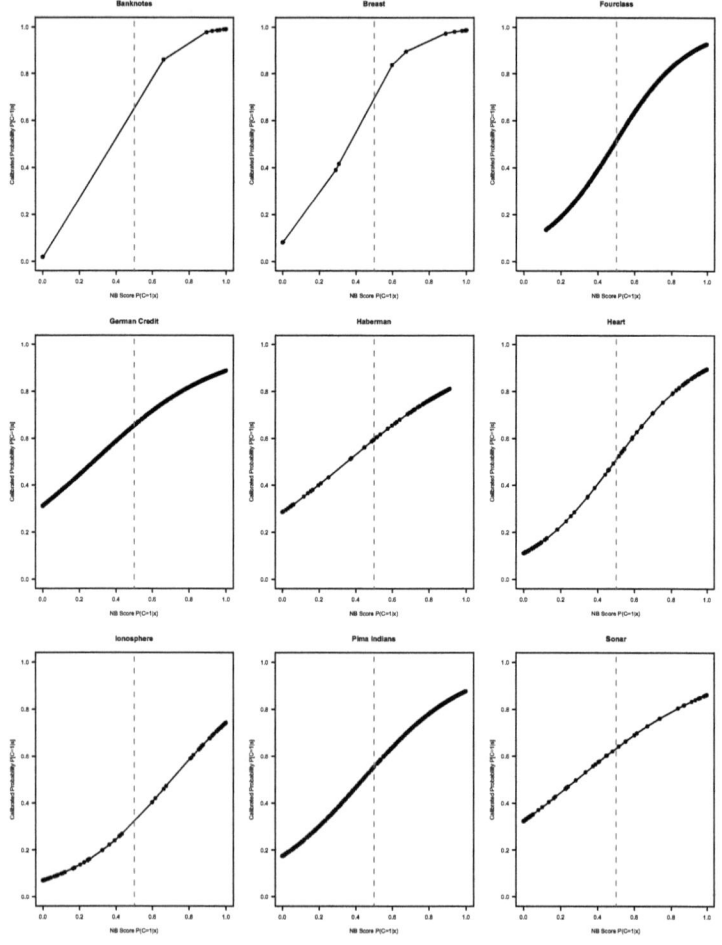

Figure A.1.: Logistic Regression on Naive Bayes Scores for two–class data sets

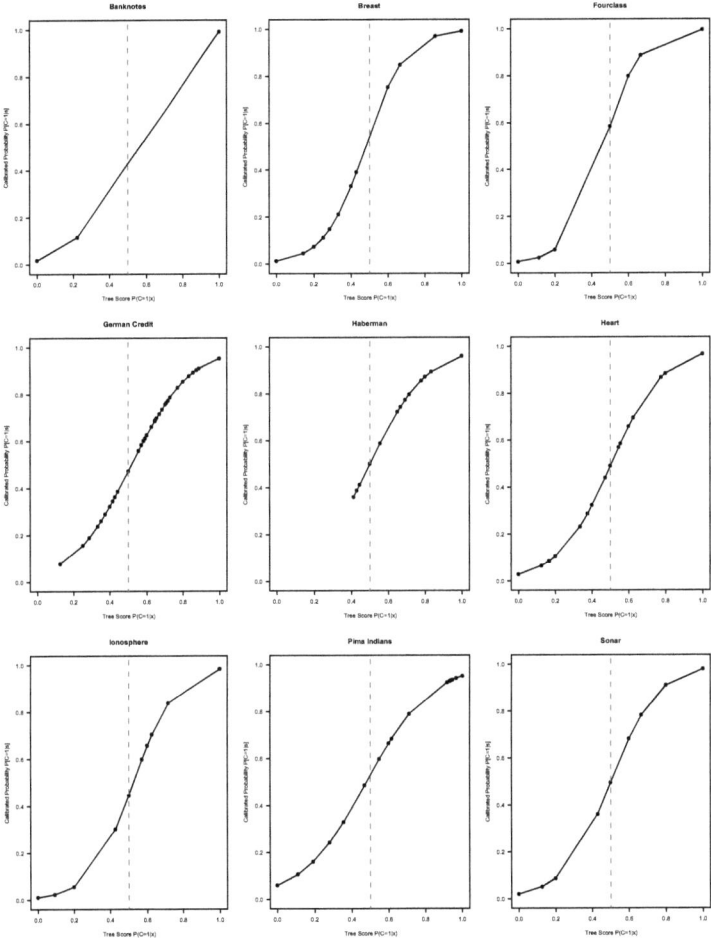

Figure A.2.: Logistic Regression on Tree Scores for two-class data sets

Table A.1.: Performance for the $L2$–SVM using one–against rest – all calibrators

	B3		Balance		Ecoli	
	CR	Cal	CR	Cal	CR	Cal
$P_{L2\text{-}SVM}(C\|\vec{x})$	0.713	0.569	0.892	0.551	0.833	0.727
$P_{av}(C\|\vec{x})$	0.694	0.590	0.894	0.754	0.791	0.784
$P_{bay\text{-}alap}(C\|\vec{x})$	0.585	0.523	0.896	0.774	0.684	0.686
$P_{ir}(C\|\vec{x})$	0.388	0.568	0.828	0.837	0.610	0.720
$P_{lr}(C\|\vec{x})$	0.687	0.558	0.892	0.831	0.824	0.736
$P_{plr}(C\|\vec{x})$	0.503	0.557	0.865	0.865	0.675	0.693
$P_{diri\text{-}g\text{-}mean}(C\|\vec{x})$	0.694	0.708	0.892	0.795	0.839	0.856
$P_{diri\text{-}g\text{-}trim}(C\|\vec{x})$	0.694	0.708	0.892	0.795	0.839	0.856
$P_{diri\text{-}prod}(C\|\vec{x})$	0.675	0.679	0.894	0.796	0.827	0.830
	Glass		Iris		Segment	
	CR	Cal	CR	Cal	CR	Cal
$P_{L2\text{-}SVM}(C\|\vec{x})$	0.481	0.625	0.926	0.560	0.838	0.666
$P_{av}(C\|\vec{x})$	0.453	0.632	0.893	0.739	0.822	0.755
$P_{bay\text{-}alap}(C\|\vec{x})$	0.378	0.607	0.793	0.625	0.634	0.706
$P_{ir}(C\|\vec{x})$	0.252	0.573	0.393	0.550	0.447	0.735
$P_{lr}(C\|\vec{x})$	0.476	0.583	0.920	0.673	0.796	0.723
$P_{plr}(C\|\vec{x})$	0.242	0.534	0.800	0.614	0.702	0.720
$P_{diri\text{-}g\text{-}mean}(C\|\vec{x})$	0.476	0.647	0.913	0.815	0.857	0.870
$P_{diri\text{-}g\text{-}trim}(C\|\vec{x})$	0.476	0.647	0.913	0.815	0.857	0.870
$P_{diri\text{-}prod}(C\|\vec{x})$	0.467	0.677	0.920	0.790	0.858	0.861
	Vehicle		Wine		Zoo	
	CR	Cal	CR	Cal	CR	Cal
$P_{L2\text{-}SVM}(C\|\vec{x})$	0.686	0.563	0.966	0.524	0.910	0.737
$P_{av}(C\|\vec{x})$	0.695	0.635	0.966	0.786	0.891	0.831
$P_{bay\text{-}alap}(C\|\vec{x})$	0.562	0.617	0.926	0.662	0.782	0.815
$P_{ir}(C\|\vec{x})$	0.492	0.630	0.668	0.692	0.752	0.808
$P_{lr}(C\|\vec{x})$	0.667	0.581	0.966	0.710	0.910	0.791
$P_{plr}(C\|\vec{x})$	0.604	0.577	0.792	0.779	0.811	0.819
$P_{diri\text{-}g\text{-}mean}(C\|\vec{x})$	0.700	0.717	0.971	0.850	0.920	0.878
$P_{diri\text{-}g\text{-}trim}(C\|\vec{x})$	0.700	0.717	0.971	0.850	0.920	0.878
$P_{diri\text{-}prod}(C\|\vec{x})$	0.703	0.706	0.971	0.811	0.891	0.892

Table A.2.: Performance for the $L2$–SVM using all–pairs – all calibrators

	B3		*Balance*		*Ecoli*	
	CR	Cal	CR	Cal	CR	Cal
$P_{L2\text{-SVM}}(C|\vec{x})$	0.738	0.531	0.873	0.514	0.824	0.646
$P_{\text{av}}(C|\vec{x})$	0.694	0.672	0.889	0.781	0.842	0.808
$P_{\text{bay-alap}}(C|\vec{x})$	0.630	0.666	0.883	0.813	0.684	0.729
$P_{\text{ir}}(C|\vec{x})$	0.573	0.581	0.870	0.853	0.794	0.759
$P_{\text{lr}}(C|\vec{x})$	0.668	0.652	0.889	0.853	0.815	0.799
$P_{\text{plr}}(C|\vec{x})$	0.592	0.612	0.864	0.732	0.818	0.807
$P_{\text{diri-g-mean}}(C|\vec{x})$	0.694	0.696	0.888	0.741	0.794	0.724
$P_{\text{diri-g-trim}}(C|\vec{x})$	0.694	0.696	0.888	0.741	0.794	0.724
$P_{\text{diri-prod}}(C|\vec{x})$	0.636	0.617	0.886	0.837	0.488	0.725
	Glass		*Iris*		*Segment*	
	CR	Cal	CR	Cal	CR	Cal
$P_{L2\text{-SVM}}(C|\vec{x})$	0.528	0.599	0.893	0.524	0.926	0.603
$P_{\text{av}}(C|\vec{x})$	0.630	0.686	0.94	0.734	0.935	0.833
$P_{\text{bay-alap}}(C|\vec{x})$	0.528	0.687	0.926	0.837	0.858	0.863
$P_{\text{ir}}(C|\vec{x})$	0.401	0.614	0.713	0.647	0.782	0.825
$P_{\text{lr}}(C|\vec{x})$	0.551	0.674	0.966	0.726	0.933	0.879
$P_{\text{plr}}(C|\vec{x})$	0.528	0.651	0.913	0.860	0.869	0.865
$P_{\text{diri-g-mean}}(C|\vec{x})$	0.570	0.676	0.880	0.690	0.940	0.822
$P_{\text{diri-g-trim}}(C|\vec{x})$	0.570	0.676	0.920	0.707	0.940	0.822
$P_{\text{diri-prod}}(C|\vec{x})$	0.476	0.662	0.893	0.847	0.888	0.843
	Vehicle		*Wine*		*Zoo*	
	CR	Cal	CR	Cal	CR	Cal
$P_{L2\text{-SVM}}(C|\vec{x})$	0.703	0.561	0.971	0.476	0.960	0.622
$P_{\text{av}}(C|\vec{x})$	0.699	0.676	0.960	0.787	0.940	0.808
$P_{\text{bay-alap}}(C|\vec{x})$	0.671	0.702	0.780	0.711	0.920	0.898
$P_{\text{ir}}(C|\vec{x})$	0.622	0.653	0.887	0.834	0.811	0.846
$P_{\text{lr}}(C|\vec{x})$	0.634	0.660	0.938	0.749	0.920	0.740
$P_{\text{plr}}(C|\vec{x})$	0.608	0.649	0.859	0.818	0.920	0.890
$P_{\text{diri-g-mean}}(C|\vec{x})$	0.691	0.673	0.949	0.745	0.950	0.825
$P_{\text{diri-g-trim}}(C|\vec{x})$	0.691	0.673	0.949	0.745	0.950	0.825
$P_{\text{diri-prod}}(C|\vec{x})$	0.692	0.696	0.926	0.850	0.861	0.907

Table A.3.: Performance for the ANN using one–against rest – all calibrators

	B3		Balance		Ecoli	
	CR	Cal	CR	Cal	CR	Cal
$P_{\text{ANN-direct}}(C|\vec{x})$	0.700	0.727	0.982	0.893	0.860	0.845
$P_{\text{ANN-rest}}(C|\vec{x})$	0.656	0.687	0.953	0.638	0.869	0.794
$P_{\text{av}}(C|\vec{x})$	0.649	0.683	0.956	0.895	0.863	0.868
$P_{\text{bay-alap}}(C|\vec{x})$	0.592	0.672	0.940	0.910	0.863	0.874
$P_{\text{ir}}(C|\vec{x})$	0.541	0.660	0.953	0.924	0.869	0.867
$P_{\text{lr}}(C|\vec{x})$	0.675	0.696	0.958	0.898	0.869	0.874
$P_{\text{plr}}(C|\vec{x})$	0.522	0.609	0.950	0.924	0.869	0.878
$P_{\text{diri-g-mean}}(C|\vec{x})$	0.643	0.663	0.918	0.853	0.866	0.875
$P_{\text{diri-g-trim}}(C|\vec{x})$	0.643	0.663	0.918	0.853	0.866	0.875
$P_{\text{diri-prod}}(C|\vec{x})$	0.662	0.687	0.926	0.876	0.526	0.713
$P_{\text{diri-direct}}(C|\vec{x})$	0.700	0.719	0.982	0.748	0.854	0.814

	Glass		Iris		Segment	
	CR	Cal	CR	Cal	CR	Cal
$P_{\text{ANN-direct}}(C|\vec{x})$	0.598	0.680	0.966	0.881	0.654	0.758
$P_{\text{ANN-rest}}(C|\vec{x})$	0.649	0.678	0.973	0.810	0.909	0.725
$P_{\text{av}}(C|\vec{x})$	0.658	0.710	0.966	0.948	0.923	0.896
$P_{\text{bay-alap}}(C|\vec{x})$	0.640	0.715	0.973	0.959	0.909	0.915
$P_{\text{ir}}(C|\vec{x})$	0.635	0.720	0.966	0.956	0.922	0.902
$P_{\text{lr}}(C|\vec{x})$	0.635	0.709	0.973	0.940	0.927	0.900
$P_{\text{plr}}(C|\vec{x})$	0.630	0.718	0.973	0.949	0.924	0.897
$P_{\text{diri-g-mean}}(C|\vec{x})$	0.649	0.709	0.973	0.958	0.909	0.934
$P_{\text{diri-g-trim}}(C|\vec{x})$	0.649	0.709	0.973	0.958	0.909	0.934
$P_{\text{diri-prod}}(C|\vec{x})$	0.579	0.692	0.973	0.934	0.912	0.934
$P_{\text{diri-direct}}(C|\vec{x})$	0.574	0.661	0.966	0.836	0.637	0.754

	Vehicle		Wine		Zoo	
	CR	Cal	CR	Cal	CR	Cal
$P_{\text{ANN-direct}}(C|\vec{x})$	0.695	0.708	0.887	0.760	0.950	0.868
$P_{\text{ANN-rest}}(C|\vec{x})$	0.743	0.632	0.960	0.788	0.970	0.898
$P_{\text{av}}(C|\vec{x})$	0.737	0.745	0.966	0.932	0.960	0.941
$P_{\text{bay-alap}}(C|\vec{x})$	0.739	0.754	0.955	0.943	0.960	0.945
$P_{\text{ir}}(C|\vec{x})$	0.738	0.750	0.949	0.936	0.861	0.903
$P_{\text{lr}}(C|\vec{x})$	0.732	0.751	0.955	0.927	0.960	0.908
$P_{\text{plr}}(C|\vec{x})$	0.737	0.750	0.926	0.924	0.940	0.908
$P_{\text{diri-g-mean}}(C|\vec{x})$	0.737	0.744	0.966	0.949	0.950	0.949
$P_{\text{diri-g-trim}}(C|\vec{x})$	0.737	0.744	0.966	0.949	0.950	0.949
$P_{\text{diri-prod}}(C|\vec{x})$	0.739	0.741	0.960	0.929	0.960	0.814
$P_{\text{diri-direct}}(C|\vec{x})$	0.679	0.702	0.893	0.845	0.950	0.908

Table A.4.: Performance for the ANN using all–pairs – all calibrators

	B3		Balance		Ecoli	
	CR	Cal	CR	Cal	CR	Cal
$P_{\text{ANN-direct}}(C|\vec{x})$	0.700	0.727	0.982	0.893	0.860	0.845
$P_{\text{ANN-pairs}}(C|\vec{x})$	0.719	0.700	0.910	0.558	0.848	0.705
$P_{\text{av}}(C|\vec{x})$	0.707	0.736	0.974	0.932	0.877	0.884
$P_{\text{bay-alap}}(C|\vec{x})$	0.687	0.691	0.966	0.929	0.860	0.871
$P_{\text{ir}}(C|\vec{x})$	0.445	0.527	0.969	0.939	0.854	0.886
$P_{\text{lr}}(C|\vec{x})$	0.700	0.729	0.972	0.937	0.863	0.853
$P_{\text{plr}}(C|\vec{x})$	0.566	0.626	0.966	0.938	0.839	0.860
$P_{\text{diri-g-mean}}(C|\vec{x})$	0.700	0.695	0.902	0.741	0.514	0.650
$P_{\text{diri-g-trim}}(C|\vec{x})$	0.700	0.695	0.902	0.741	0.514	0.650
$P_{\text{diri-prod}}(C|\vec{x})$	0.726	0.729	0.945	0.889	0.836	0.885
$P_{\text{diri-direct}}(C|\vec{x})$	0.700	0.719	0.982	0.748	0.854	0.814
	Glass		Iris		Segment	
	CR	Cal	CR	Cal	CR	Cal
$P_{\text{ANN-direct}}(C|\vec{x})$	0.598	0.680	0.966	0.881	0.654	0.758
$P_{\text{ANN-pairs}}(C|\vec{x})$	0.686	0.659	0.973	0.780	0.916	0.701
$P_{\text{av}}(C|\vec{x})$	0.686	0.733	0.973	0.945	0.950	0.942
$P_{\text{bay-alap}}(C|\vec{x})$	0.635	0.717	0.973	0.964	0.943	0.944
$P_{\text{ir}}(C|\vec{x})$	0.640	0.716	0.966	0.951	0.945	0.947
$P_{\text{lr}}(C|\vec{x})$	0.691	0.720	0.973	0.907	0.950	0.940
$P_{\text{plr}}(C|\vec{x})$	0.630	0.720	0.966	0.946	0.949	0.944
$P_{\text{diri-g-mean}}(C|\vec{x})$	0.630	0.700	0.973	0.817	0.942	0.801
$P_{\text{diri-g-trim}}(C|\vec{x})$	0.630	0.700	0.973	0.817	0.942	0.801
$P_{\text{diri-prod}}(C|\vec{x})$	0.677	0.730	0.973	0.941	0.918	0.938
$P_{\text{diri-direct}}(C|\vec{x})$	0.574	0.661	0.966	0.836	0.637	0.754
	Vehicle		Wine		Zoo	
	CR	Cal	CR	Cal	CR	Cal
$P_{\text{ANN-direct}}(C|\vec{x})$	0.695	0.708	0.887	0.760	0.950	0.868
$P_{\text{ANN-pairs}}(C|\vec{x})$	0.757	0.585	0.943	0.781	0.910	0.786
$P_{\text{av}}(C|\vec{x})$	0.755	0.769	0.943	0.929	0.930	0.919
$P_{\text{bay-alap}}(C|\vec{x})$	0.748	0.769	0.938	0.912	0.950	0.922
$P_{\text{ir}}(C|\vec{x})$	0.751	0.777	0.949	0.926	0.801	0.804
$P_{\text{lr}}(C|\vec{x})$	0.756	0.770	0.943	0.920	0.930	0.764
$P_{\text{plr}}(C|\vec{x})$	0.752	0.772	0.932	0.910	0.871	0.860
$P_{\text{diri-g-mean}}(C|\vec{x})$	0.754	0.759	0.943	0.826	0.960	0.719
$P_{\text{diri-g-trim}}(C|\vec{x})$	0.754	0.759	0.943	0.826	0.960	0.719
$P_{\text{diri-prod}}(C|\vec{x})$	0.749	0.727	0.943	0.923	0.940	0.919
$P_{\text{diri-direct}}(C|\vec{x})$	0.679	0.702	0.893	0.845	0.950	0.908

Die VDM Verlagsservicegesellschaft sucht für wissenschaftliche Verlage abgeschlossene und herausragende

Dissertationen, Habilitationen, Diplomarbeiten, Master Theses, Magisterarbeiten usw.

für die kostenlose Publikation als Fachbuch.

Sie verfügen über eine Arbeit, die hohen inhaltlichen und formalen Ansprüchen genügt, und haben Interesse an einer honorarvergüteten Publikation?

Dann senden Sie bitte erste Informationen über sich und Ihre Arbeit per Email an *info@vdm-vsg.de*.

Sie erhalten kurzfristig unser Feedback!

VDM Verlagsservicegesellschaft mbH
Dudweiler Landstr. 99 Telefon +49 681 3720 174
D - 66123 Saarbrücken Fax   +49 681 3720 1749
www.vdm-vsg.de

Die VDM Verlagsservicegesellschaft mbH vertritt

Printed by Books on Demand GmbH, Norderstedt / Germany